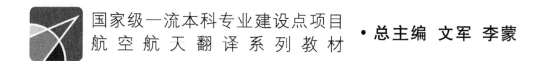

国家级一流本科专业建设点项目
航空航天翻译系列教材 ・总主编 文军 李蒙

航空航天文学翻译教程

A Coursebook on Translation of Aeronautic and Astronautic Literature

主　编　武光军
副主编　王佳敏　陈晓璐

北京航空航天大学出版社

内 容 简 介

本书是航空航天文学翻译教材。它突出理论与实践相结合的理念,在系统阐述航空航天文学翻译原则的基础上,着力介绍了航空航天文学翻译的策略与方法,特别是从英译汉和汉译英两个方向阐述了航空航天文学翻译的原则与方法,介绍了航空航天文学翻译批评与赏析的基本模式,并在每节的最后都配有相关练习,便于学生掌握相关翻译策略及翻译方法。教材还精选了大量第一手的航空航天文学翻译素材,以帮助学生提高航空航天文学翻译的综合能力。

本教材适用于翻译专业本科生及翻译硕士。

图书在版编目(CIP)数据

航空航天文学翻译教程 / 武光军主编. -- 北京：北京航空航天大学出版社,2024.7. -- ISBN 978-7-5124-4443-0

Ⅰ.Ⅴ

中国国家版本馆 CIP 数据核字第 2024EG1866 号

版权所有,侵权必究。

航空航天文学翻译教程

主编　武光军

副主编　王佳敏　陈晓璐

策划编辑　赵延永　蔡　喆

责任编辑　张　凌

*

北京航空航天大学出版社出版发行

北京市海淀区学院路 37 号(邮编 100191)　http://www.buaapress.com.cn

发行部电话：(010)82317024　传真：(010)82328026

读者信箱：goodtextbook@126.com　邮购电话：(010)82316936

北京九州迅驰传媒文化有限公司印装　各地书店经销

*

开本：710×1 000　1/16　印张：10　字数：213 千字

2024 年 7 月第 1 版　2024 年 7 月第 1 次印刷

ISBN 978-7-5124-4443-0　定价：35.00 元

若本书有倒页、脱页、缺页等印装质量问题,请与本社发行部联系调换。联系电话：(010)82317024

编委会

总 主 编 文 军 李 蒙

编　　委（按姓氏笔画排序）

　　　　　　孔　杰　朱殿勇　许明武　李　雪

　　　　　　李京廉　张化丽　张艳玲　范祥涛

　　　　　　赵雪琴　秦明利　梁茂成　彭　静

丛书策划 赵延永　蔡　喆

本书主编 武光军

本书副主编 王佳敏　陈晓璐

总序
Foreword

科学技术的发展离不开交流与合作,航空航天的发展也不例外。在中国航空航天发展史上,这种交流与合作很大程度上与翻译相关,概括起来,大致经历了两个阶段:早期的翻译引进,后期的翻译引进与输出并举。最早与航空航天有关的翻译引进活动始于1903年到1907年期间中国掀起的"凡尔纳热",其中的航空航天科幻小说翻译包括鲁迅的《月界旅行》、商务印书馆出版的《环游月球》以及谢祺的《飞行记》等。1910年高鲁翻译出版了《空中航行术》,这是中国航空航天科技书籍和资料汉译的开端。而在译出方面,随着我国航空航天事业的飞速发展,近些年的科技新闻、政府白皮书等都有大量航空航天方面的信息对外发布,及时而系统地向全世界展现了中国在此领域的发展现状和巨大成就。

总体而言,航空航天的领域宽广,翻译多种多样。从翻译的主题看,航空航天话语以科技语言为主,其一般特点有七个方面:无人称、语气正式、陈述客观准确、语言规范、文体质朴、逻辑性强和专业术语性强[1],与之相关的科技论文等翻译是航空航天翻译的主体。此外,航空航天话语中还包括与商务活动相关的商贸翻译(如合同、谈判等)、与航空航天新闻活动相关的新闻翻译(如新闻发布会、各种媒体的相关新闻报道等)、与航空航天文学相关的文学翻译(如航空类小说、航天类科幻小说等)、与航空航天影视活动相关的影视翻译(如纪录片、科幻电影)等;从翻译活动的方式看,航空航天翻译包括了笔译、视译、交替传译、同声传译、机器翻译+译后编辑等几乎所有翻译方式。

航空航天翻译主题和体裁的多样性及翻译方式的全面性,对翻译人才的培养提出了新的、更高的要求。为此,我们特设计和编写了这套"航空航天翻译系列教材",其特色主要体现在以下几个方面:

1."入主流"与"显特色"并举。"入主流"主要指各种教材的设计都体现了翻译这一核心要素,其内容选择都以"怎么翻译"为焦点;"显特色"则体现在教材内容的选

---

[1] 冯志杰.汉英科技翻译指要[M].北京:中国对外翻译出版公司,1998:6-7.

择上，无论是例句还是练习，都择选了与航空航天密切相关的语料，力求解决航空航天翻译中的实际问题。

2. 理论与实践并重。在教材设计上，突显理论融于实践的理念，对理论不做大篇幅的阐释，而将翻译策略、翻译方法等融于对例句和语篇的解释之中，而这些例句和语篇都选自真实的航空航天语料，以着力提升学生的翻译实践能力。

3. 阐释与练习并立。对各种翻译现象的解释与阐释在教材中必不可少，是教材的主干；与此同时，各教材采用按"节"的方式设置练习，其用意在于着力加强练习与教材正文的关联性，以方便学生的学习和操练。

本系列教材可以作为翻译专业、英语专业和大学英语相关课程的课堂教学材料，也可供对航空航天翻译感兴趣的读者使用。

迄今，本系列教材已规划了英汉翻译、汉英翻译、口译、影视翻译等教材；今后，我们还可增加与航空航天翻译相关的品种，如航空航天文学翻译、航空航天新闻翻译、商贸翻译、航空航天同声传译等教材。

为使本系列教材的编写更具广泛性和权威性，我们组建了高水平的编委会。编委会委员有北京航空航天大学文军、李蒙，北京理工大学李京廉，重庆大学彭静，大连理工大学秦明利，哈尔滨工业大学李雪，哈尔滨工程大学朱殿勇，华中科技大学许明武，南京航空航天大学范祥涛，南京理工大学赵雪琴，西北工业大学孔杰，西安航空学院张化丽和中国民航大学张艳玲等专家学者。

本系列教材的编写是一种尝试，希望得到业内专家学者、学生和读者的反馈及意见，以使教材更臻完善。

文军　李蒙

2023 年 3 月于北京

前言 Preface

本教材突出理论与实践相结合的理念,在系统阐述航空航天文学翻译原则的基础上,着力介绍了航空航天文学翻译的策略与方法,特别是从英译汉和汉译英两个方向阐述了航空航天文学翻译的原则与方法,介绍了航空航天文学翻译批评与赏析的基本模式,并在每节的最后都配有相关练习,便于学生掌握相关翻译策略及翻译方法。教材还精选了大量第一手的航空航天文学翻译素材,以帮助学生提高航空航天文学翻译的综合能力。

作为国内尚不多见的以航空航天文学翻译为主题的教材,本书的编写特色可概括为以下三点:

一是语料的第一手性。在航空航天事业不断发展的进程中,文学领域也产生了众多优秀的航空航天文学作品。要成为航空航天领域的强国,我们一方面需要引进全世界的优秀航空航天文学作品,另一方面也要将我国优秀的航空航天文学作品推广到全世界。本教材中的语料选自当代第一手的航空航天文学翻译文本,意在让学生接触第一手语料,熟悉相关文体及其译文,进而提高翻译能力。

二是翻译策略与方法的实用性。本教材不仅在第二章介绍了航空航天文学翻译的基本原则,还在第三章和第四章较为系统地介绍了常用的翻译策略与方法:翻译策略分为异化翻译策略与归化翻译策略两大方面;在翻译方法方面,在异化翻译策略下具体分析了借用(borrowing)、仿造(calque)、直译(literal translation)的翻译方法,在归化翻译策略下具体分析了置换(transposition)、调适(modulation)、对等(equivalence)、显化(explication)、隐化(implicitation)、改编(adaptation)的翻译方法。这些翻译策略与方法形成了一个覆盖全面的系统,具有较强的实用性,解释力较强,便于学生掌握、运用及解决航空航天文学翻译问题。

三是练习的拓展性。本教材中的练习按节设计,目的是确保练习与所学内容的相关性,同时还设置了对学习内容进行拓展的练习——我们在每节的练习中都设置了"基础练习"和"拓展练习"两个板块,"基础练习"侧重于对该节内容的理解与操练,"拓展练习"则关注知识与技能的延展。

上述特点保证了本教材的专业性、实用性和灵活性。随着我国航空航天事业的飞速发展,航空航天文学翻译必将迎来大发展的机遇期。本教材旨在为学生提供入门指导,为他们提高航空航天文学翻译能力提供帮助,同时也为我国航空航天走向世界助一臂之力。

武光军
2023 年 6 月于北航

目　　录

第一章　航空航天文学翻译概述 …………………………………… 1

第一节　文学文本的语言特征 …………………………………… 2
第二节　航空航天文学文本的语言特征 ………………………… 7
第三节　航空航天文学的翻译概况 ……………………………… 14

第二章　航空航天文学翻译原则 …………………………………… 17

第一节　文学翻译原则 …………………………………………… 18
第二节　航空航天文学翻译：英译汉的原则 …………………… 23
第三节　航空航天文学翻译：汉译英的原则 …………………… 31

第三章　航空航天文学翻译的策略与方法：英译汉 ……………… 43

第一节　航空航天文学翻译英译汉的策略 ……………………… 43
第二节　航空航天文学翻译英译汉的方法 ……………………… 48

第四章　航空航天文学翻译的策略与方法：汉译英 ……………… 72

第一节　航空航天文学翻译汉译英的策略 ……………………… 72
第二节　航空航天文学翻译汉译英的方法 ……………………… 74

第五章　航空航天文学翻译批评与赏析 …………………………… 88

第一节　文学翻译批评与赏析 …………………………………… 88
第二节　航空航天文学翻译批评与赏析：英译汉 ……………… 100
第三节　航空航天文学翻译批评与赏析：汉译英 ……………… 106

第六章　航空航天文学翻译实践 …………………………………………… 111

第一节　航空航天文学翻译实践：英译汉 ……………………………… 111
第二节　航空航天文学翻译实践：汉译英 ……………………………… 115

练习题及翻译实践的参考答案 ……………………………………………… 119

第一章 …………………………………………………………………… 119
第二章 …………………………………………………………………… 123
第三章 …………………………………………………………………… 125
第四章 …………………………………………………………………… 128
第五章 …………………………………………………………………… 131
第六章 …………………………………………………………………… 137

第一章　航空航天文学翻译概述

当今,文化在综合国力竞争中的地位和作用越来越受到各国的重视。在我国,文化越来越从边缘走向国家政策的中心,文化大发展也被提升到国家战略高度。2006年发布的《国家"十一五"时期文化发展规划纲要》把文化"走出去"放到了一个重要的战略位置,从"拓展对外文化交流和传播渠道""培育外向型骨干文化企业""实施'走出去'重大工程项目"三大方面系统规划了文化走出去工程,使文化"走出去"有了具体的路径[①]。2012年11月,党的十八大把文化建设作为与经济建设、政治建设、社会建设、生态文明建设相并列的治国策略,文化建设正式上升到国家战略层面[②]。特别是文化"走出去",对增强我国国家文化软实力、扩大国际影响力意义重大。约瑟夫·乃[③]指出,一国的软实力主要来源于三个方面:该国文化所传达的价值观、该国的实践及政策所树立的典范以及该国处理与他国关系的方式。由此可见,软实力的第一大来源就是文化。因此,我们要从国家战略高度思考我国的文化"走出去"。2013年8月,中央宣传工作会议在北京召开,习近平总书记作了重要讲话,强调要"创新对外宣传方式,着力打造融通中外的新概念新范畴新表述,讲好中国故事,传播好中国声音"。

要讲好中国故事,传播好中国声音,自然就要求翻译发挥越来越重要的作用。在当前中国文化"走出去"的国家战略中,文学翻译已成为中国增强软实力的战略性方式。文学翻译中,特别是中国文学英译的出版与传播是我国文化"走出去"战略的重要一环[④]。这是因为,当今英语仍是国际影响力最大的语言,要想扩大国际影响力,就不得不译入英语。同时,翻译作品在英美图书市场的接受度极低。在英国或美国,翻译书目仅占其图书市场的2%~4%,远低于法国的15%~18%,德国的11%~14%,意大利的25%,西班牙的25%~26%[⑤]。从接受度而言,较其他语种,中国文学英译的难度要大很多。因此,中国文学英译在我国当下的翻译研究中具有重要地位。

航空航天是当今世界最具挑战性和广泛带动性的高科技领域之一,航空航天活动深刻改变了人类对宇宙的认知,为人类社会进步提供了重要动力。我国把发展航

[①] 王廷信.中国艺术海外传播的国家战略与理论研究[J].民族艺术,2017(2):20-27.
[②] 同①.
[③] Nye Jr J. Soft Power: The Means to Success in World Politics[M]. New York: Public Affairs, 2004.
[④] 王建开.走出去战略与出版意图的契合:以英译作品的当代转向为例[J].上海翻译,2014(4):1-7.
[⑤] Ganne V, Minon M. Géographies de la traduction[M]//Françoise Barret Ducrocq. Traduire l'Europe. Paris: Payot, 1992: 55-95.

空航天事业作为国家整体发展战略的重要组成部分,始终坚持为和平目的探索和利用外层空间。我国的航空航天事业创建于1956年,至今已走过60多年的发展历程,历经1956—1978年的艰苦创业、1978—1999年的改革振兴、1999年至今的跨越发展三个阶段。经过几代航天人的接续奋斗,创造了以人造卫星、载人航天、月球探测、火星探测等里程碑为代表的辉煌成就,走出了一条自力更生、自主创新的发展道路,实现了从"跟跑"到"并跑"和"部分领跑"、从传统模式向现代模式、从试验型应用向业务化应用、从服务国内到造福人类的"四个转变"。我国航空航天科技创新能力显著增强,空间应用服务能力大幅提升,开放合作深化拓展,已跻身于世界航空航天强国的行列。在航空航天事业不断发展的进程中,文学领域也产生了众多优秀的航空航天文学作品。要成为航空航天领域的强国,我们一方面需要引进全世界的优秀航空航天文学作品,另一方面也要将我国优秀的航空航天文学作品推广到全世界。

第一节　文学文本的语言特征

文学是语言的艺术,语言是文学的载体。由此可见,文学对语言有着特殊的要求。简言之,与普通语言注重传达信息不同,文学语言强调审美。文学语言与普通语言最大的区别就是,普通语言是所指优势符号,而文学语言是能指优势符号。所指优势符号即普通语言使解释者按照严格的规定来理解,尽量排斥不同的解释,使编码具有一定的权威性,而能指优势符号即文学语言具有返回能指性、创新性[1],能指与所指的关系具有单一性和透明性。陈学广[2]指出,理想的科学语言是一种抽象的单义性语言,它要求语义固定,一词一义;而文学语言则是一种意象化的多义性语言,它所使用的语词"必须包孕各种意义",具有"潜在意义能力",从而使语言在文学语境中派生出更多的潜在意义。布鲁克斯[3]在《释义误说》中比较科学语言与诗歌语言的语义差异时指出:"科学的术语是抽象的符号,它们不会在语境的压力下改变意义。它们是纯粹的(或者说渴望它们是纯粹的)语义;它们是事先就被限定好的。它们不会被歪曲到新的语义之中。可是哪儿有能包含一首诗的用语的辞典呢?诗人被迫不断地再创造语言,这已是一句老生常谈。正如艾略特所说,他的任务就是'使语言脱臼进入意义'。而且从科学词汇的视角看,这正是诗人所起的作用;因为从推理上考虑,理想的语言应该是一词一义,并且词和义之间的关系也应该是稳定的。但诗人使用的词却必须包孕各种意义,不是不连续的意义碎片,而是有潜在意义能力的词,即意

---

[1]　龚艳. 文学语言特征的美学探析[J]. 传奇·传记文学选刊, 2011(4):13-14.
[2]　陈学广. 文学语言:直接意指与含蓄意指:文学语义系统及其特征解析[J]. 江苏社会科学, 2007(1): 199-203.
[3]　克利安思·布鲁克斯. 释义误说[M]. 杜定宇, 译//赵毅衡. "新批评"文集. 天津:百花文艺出版社, 2001.

的网络或意义的集束。"以审美化的语言形式作为塑造艺术形象的手段和工具的文学活动,又始终不能离开创造、创新,它所超越于科学语言和日常语言的审美化的创新特征正是文学的形象思维具有独光异彩的主要表现①。整体来说,文学语言往往具有如下特征:形象性、凝练性、陌生化。

一、形象性

文学语言具有塑造文学形象的功能,因此形象性是文学语言的一大特征。文学作品越伟大,其语言的形象性往往就越突出。例如,朱自清的《荷塘月色》就为读者描绘出了令人难忘的"月光"形象:

例1 中文原文:

月光如流水一般,静静地泻在这一片叶子和花上。薄薄的青雾浮起在荷塘里。叶子和花仿佛在牛乳中洗过一样;又像笼着轻纱的梦。

英文译文一:

The moon's rays were like flowing waters, gently depositing their moisture on the layer of leaves and blossoms. A light green mist floated just above the lotus pond. The leaves and blossoms looked as though they had been bathed in milk, or like a blurred dream swathed in airy gauze.(葛浩文译)

英文译文二:

Moonlight cascaded like water over the lotus leaves and flowers, and a light blue mist floating up from the pool made them seem washed in milk or caught in a gauzy dream.(杨宪益、戴乃迭译)

杨守国②对此进行了详细的解读:"流水"般的月光倾泻,薄薄的青雾浮起,这是实写;而"仿佛在牛乳中洗过一样""又像笼着轻纱的梦"则是虚拟,但虚中见实,贴切地摹状了朦胧月色下荷花飘忽的姿态。虚实为用,勾勒出一种摄人心魄的意境。"流水""牛乳""轻纱"三个比喻都写月色的美好,但"流水"重在喻其柔和,"牛乳"重在喻其"洁白","轻纱"则重在喻其淡微,比喻呈现出面与面的扩展,整合为一,则形神兼备,形容尽致。

葛浩文的译文及杨宪益、戴乃迭的译文都注意到了朱自清原文中的形象性,都注意在英语译文中保留这种形象性。整体来说,葛浩文的译文更忠实于原文,塑造的月光的形象更完备。例如,原文中"静静地泻在这一片叶子和花上"这句的确不好译,因修饰动词"泻"的副词"静静地"似乎与"泻"在语义上是矛盾的。鉴于此,杨宪益、戴乃迭将此部分省略不译,但葛浩文的译文进行了变通处理,译为了"gently depositing

---

① 杨守国. 文学语言的艺术功能与创新特征:以朱自清《荷塘月色》为例[J]. 西安石油大学学报,2009 (2):103-106.

② 同①。

their moisture on the layer of leaves and blossoms".

二、凝练性

凝练性指的是文学语言来自日常语言,但往往又高于日常语言。文学语言是高度凝练、高度浓缩化的艺术性语言。文学语言的凝练性往往会要求俗语、成语、修辞等各种手段的运用。例如,余华的《许三观卖血记》中有一段话,讲述的是二乐的队长去许三观家和他喝酒:

例 2　中文原文:

"不行,"二乐的队长说,"要全喝了,这叫感情深,一口吞;感情浅,舔一舔。"①

英文译文:

"That won't do," the brigade chief said. "You've got to down it all in one gulp. That is a test of our friendship. Friends drink when they're together. Acquaintances merely sip."(安德鲁·琼斯译)

"感情深,一口吞;感情浅,舔一舔",这是中国酒文化中常说的一句话,简单明了,句式对仗,高度凝练。在此,余华采用了这句俗语,体现了其语言选用上的凝练性。琼斯的英文译文对原文中包含中国文化的语句"感情深,一口吞;感情浅,舔一舔"略作调整,添加了一个概况性语句"That is a test of our friendship",然后基本保留了后面两句的内容:Friends drink when they're together. Acquaintances merely sip。

例 3　中文原文:

曲曲折折的荷塘上面,弥望的是田田的叶子。叶子出水很高,像亭亭的舞女的裙。层层的叶子中间,零星地点缀着些白花,有袅娜地开着的,有羞涩地打着朵儿的;正如一粒粒的明珠,又如碧天里的星星,又如刚出浴的美人。(朱自清,《荷塘月色》)

英文译文一:

On the surface of the winding and twisting lotus pond floated an immense field of leaves. The leaves lay high in the water, rising up like the skirts of a dancing girl. Amid the layers of leaves white blossoms adorned the vista, some beguilingly open and others bashfully holding their petals in. Just like a string of bright pearls or stars in a blue sky, or like lovely maidens just emerging from their bath.(葛浩文译)

英文译文二:

As far as eye could see, the pool with its winding margin was covered with trim leaves, which rose high out of the water like the flared skirts of dancing girls. And starring these tiers of leaves were white lotus flowers, alluringly open or bashfully in bud, like glimmering pearls, stars in an azure

---

① 余华. 许三观卖血记[M]. 北京:作家出版社,2008.

sky, or beauties fresh from the bath.(杨宪益、戴乃迭译)

朱自清先生的这段对荷花的描写选词考究,高度凝练,如"弥望""田田的""亭亭的""袅娜地""打着朵儿",还运用了许多形象的修辞手段,如"像亭亭的舞女的裙""羞涩地打着朵儿的""如一粒粒的明珠""如碧天里的星星""如刚出浴的美人"。葛浩文的译文及杨宪益、戴乃迭的译文都较好地再现了原文的语言特征。

三、陌生化

陌生化是文学作品实现其艺术目的的一种手段,是为了让人们换一个角度,换一种眼光,换一种说法,用一种新奇的方式感受和体验事物,把人们习以为常的东西转化为一种具有新的意义新的生命力的语言感觉[①]。例如:

例 4 中文原文:

八月深秋,无边无际的高粱红成汪洋的血海,高粱高密辉煌,高粱凄婉可人,高粱爱情激荡。[②]

英文译文:

In late autumn, during the eighth lunar month, vast stretches of red sorghum shimmered like a sea of blood. Tall and dense, it reeked of glory; cold and graceful, it promised enchantment; passionate and loving, it was tumultuous.(葛浩文译)

形容词"辉煌"一般和"事业""业绩"等进行搭配,形容词"凄婉可人"和"爱情激荡"一般和指称人的词语进行搭配,但在此处,作家莫言都用来和"高粱"搭配,发生了语义范围上的偏离,属创造性搭配,即陌生化。这句话的后半部分如果按照字面直译就是:The red sorghum was tall, dense, splendid, graceful and tumultuous。在此译文中,葛浩文打破了"The red sorghum was+形容词"的简单的句法结构模式,转而使用了"形容词作状语+主句"的句法结构模式。在新的句法结构框架下,通过将三个形容词中的两个转换为名词,即将"辉煌"和"可人"转换为"glory"和"enchantment",又通过添加两个动词"reeked of"和"promised",使得整个英文译文不仅巧妙地保留了原文的创造性搭配,而且还具有较好的可读性和文学性。

例 5 中文原文:

别人都生活在土地上,生活在房屋里,我和父亲却生活在船上,这是我父亲十三年前作出的选择,他选择河流,我就只好离开土地,没什么可抱怨的。

向阳船队一年四季来往于金雀河上,所以,我和父亲的生活方式更加接近鱼类,时而顺流而下,时而逆流而上,我们的世界是一条奔涌的河流,狭窄而绵长,一滴水机械地孕育另一滴水,一秒钟沉闷地复制另一秒钟。河上十三年,我经常在船队泊岸的

---

① 田文强.文学语言陌生化的审美特征[J].湖北民族学院,2003(2):43-46.
② 莫言.红高粱家族[M].台北:洪范书店有限公司,1988.

时候回到岸上,去做陆地的客人,可是众所周知,我父亲从岸上消失很久了,他以一种草率而固执的姿态,一步一步地逃离岸上的世界,他的逃逸相当成功,河流隐匿了父亲,也改变了父亲,十三年以后,我从父亲未老先衰的身体上发现了鱼类的某些特征。(苏童,《河岸》)

英文译文:

Most people live on dry land, in houses. But my father and I live on a barge. Nothing surprising about that, since we are boat people; the terra firma does not belong to us.

Everyone knows that the Sunnyside Fleet plies the waters of the Golden Sparrow River all year round, so life for Father and me hardly differs from that of fish: whether heading upriver or down, most of our time is spent on the water. It's been thirteen years. I'm still young and strong, but my father, a rash and careless man, is sinking inexorably into the realm of the aged.(葛浩文译, *The Boat to Redemption*)

在中文原文中,苏童对"父亲"的描述采用了一种陌生化的奇特的比喻"鱼"。这是富有想象力的比喻形式,正是这种陌生化的比喻手法给读者留下了深刻的印象。但遗憾的是,译者葛浩文并不是十分注重原作者的这一陌生化的修辞手法,在译文中没有保留这些新奇的隐喻。

练 习 题

一、基础练习:请翻译以下句子

1. She had just finished setting it to rights, and was shaking her duster from the window, when she saw the two men again.①

2. They were standing now by the car and, Miss Clare was glad to see, they were doing their best to wipe the mud from their shoes on the grass.②

3. 悖时砍脑壳的!③

4. 横顺人是"牛肉炒韭菜,各人心里爱"。④

二、拓展练习:请翻译以下段落

1. "They'd got a new-fangled thing-brief-case, ain't it? —in the back. Two

---

① 宋楠.语境理论背景下文学翻译策略的选择和应用:以英语小说《乡村风暴》翻译实践为例[J].牡丹江大学学报,2015,24(05):135-138+147.
② 同①.
③ 沈从文.沈从文全集:第8卷[M].太原:北岳文艺出版社,2002.
④ 同③.

strangers, poking about here with a brief-case and a lot of mud on their shoes," she, mused. "Makes you think, don't it? Might be Ag. men, of course. But you mark my words, Miss Read, they was up to a bit of no good!"①

2. 车是车路,马是马路,各有规矩。想爸爸作主,请媒人正正经经来说是车路;要自己作主,站到对溪高崖竹林里为你唱三年六个月的歌是马路。②

第二节 航空航天文学文本的语言特征

与普通文学语言相同,航空航天文学文本的语言也往往具有形象性、凝练性与陌生化的特征。

一、形象性

与普通文本相比,文学文本的语言特征之一就是形象性。通过文学语言的描述,许多我们见过或没见过的现象都会跃然纸上。虽然我们大部分人都没坐太空飞船去太空探索过,但读了下面这段文字,相信你已经身临其境,如自己坐上了太空飞船一样:

例6 英文原文:

The Windows of a spacfship casually frame miracles. Every 92 minutes, another sunrise: a layer cake that starts with orange, then a thick wedge of blue, then the richest, darkest icing decorated with stars. The secret patterns of our planet are revealed: mountains bump up rudely from orderly plains, forests are green gashes edged with snow, rivers glint in the sunlight, twisting and turning like silvery worms. Continents splay themselves out whole, surrounded by islands sprinkled across the sea like delicate shards of shattered eggshells.③

中文译文:

飞船的舷窗随时可能向你展现奇迹。每隔92分钟就会有另一场日出:天空好像奶油蛋糕一样,最下面一层是橘色,往上是一层厚厚的蓝色,再往上是浓重的黑暗,其上繁星点点。我们所在的星球的一切秘密都展示在眼前:宽阔的平原上,山峰兀然耸立;绿色的森林边上镶嵌着皑皑白雪;阳光下,河流闪闪发光,像银色的长蛇蜿蜒伸

---

① 宋楠.语境理论背景下文学翻译策略的选择和应用:以英语小说《乡村风暴》翻译实践为例[J].牡丹江大学学报,2015,24(05):135-138+147.
② 沈从文.沈从文全集:第8卷[M].太原:北岳文艺出版社,2002.
③ Chris H. An Astronaut's Guide to Life on Earth[M]. Toronto: Random House Canada, 2013.

展;陆地完整地铺展在你的面前,周围的海洋里,有星罗棋布的岛屿,像是蛋壳的碎片,散落周围。①

下面这段文字描述的是1941年日本突袭我国香港的事件。整个事件是通过某个人的所见、所闻、所感描述出来的,可以说是感同身受,非常形象。这也正是文学的重要价值之一。

例7 英文原文:

Helen Yee Ling Chow heard the droning of planes overhead and the thud of bombs dropping. The sounds vibrated through her very bones. She could feel her brother's heartbeat pulse through his body, closely pressed to hers. Their arms were wrapped around each other. Tears ran down Helen's cheeks and pooled in the crevices of her older brother's neck. Terrified of making a noise, they felt a fear that was intensified by the silence imposed on them. In their dark hiding place, the only thing the children could hear was the war closing in all around. Hong Kong was falling.②

中文译文:

海伦·凌周叶听到头顶传来飞机的轰鸣,紧接着是炸弹落地的巨响。猛烈的震颤一直传到了她的骨头深处。她甚至能透过哥哥的身体感觉到他的心跳,他们俩挤成一团,胳膊紧紧挽在一起。泪水顺着海伦的脸颊滴落在哥哥的脖子上,他们不敢发出声音,寂静带来的恐惧压得人喘不过气来。躲在黑暗的小空间里,孩子们只能听到外面的枪炮声越来越近。香港正在沦陷。③

二、凝练性

航空航天文学文本的另一个重要特征是凝练性,即在语言运用上,航空航天文学文本的语言较为练达,高度精炼,用词较为考究,文学性较强。下面是刘慈欣对一场星际战争过程的简要描述,用词考究,写作凝练。

例8 中文原文:

在距地球五万光年的远方,在银河系的中心,一场延续了两万年的星际战争已接近尾声。

那里的太空中渐渐隐现出一个方形区域,仿佛灿烂的群星的背景被剪出一个方口,这个区域的边长约十万公里,区域的内部是一种比周围太空更黑的黑暗,让人感到一种虚空中的虚空。从这黑色的正方形中,开始浮现出一些实体,它们形状各异,都有月球大小,呈耀眼的银色。这些物体越来越多,并组成一个整齐的立方体方阵。

---

① 哈德菲尔德.宇航员地球生活指南[M].徐彬,译.长沙:湖南科学技术出版社,2017.
② Holt N. Rise of the Rocket Girls[M]. New York: Little, Brown and Company, 2016.
③ 娜塔莉娅·霍尔特.让火箭起飞的女孩[M].阳曦,译.北京:九州出版社,2022.

这银色的方阵庄严地驶出黑色正方形,两者构成了一幅挂在宇宙永恒墙壁上的镶嵌画,这幅画以绝对黑体的正方形天鹅绒为衬底,由纯净的银光耀眼的白银小构件整齐地镶嵌而成。这又仿佛是一首宇宙交响乐的固化。渐渐地,黑色的正方形消融在星空中,群星填补了它的位置,银色的方阵庄严地悬浮在群星之间。

银河系碳基联邦的星际舰队,完成了本次巡航的第一次时空跃迁。(刘慈欣,"乡村教师"《刘慈欣短篇小说合集》)

英文译文:

In the center of the Milky Way, fifty thousand light-years from Earth, an interstellar war that had lasted for twenty thousand years was nearing its resolution.

A square-shaped, starless region was visible there, as distinctly as if it had been cut from the background of shining stars with a pair of scissors. Its sides were six thousand miles long, and its interior was blacker even than the blackness of space—a void within a void. Several objects began to emerge from within the square. They were of various shapes, but each was as large as Earth's moon, and their color was a dazzling silver. As more appeared, they took on a regular, cube-shaped formation. The cube of objects continued to emerge from the square, a mosaic set into the eternal wall of the universe itself, whose base was the complete, velvet blackness of the square and whose tiles were the luminescent silver objects. They were like a cosmic symphony given physical form. Slowly, the black square dissolved back into the stars, leaving only the cube-shaped array of silver objects floating ominously.

The interstellar fleet of the Galactic Federation of Carbon-Based Life had completed the first space-time warp of its journey. (Lanphier Adam 译, *The Village Teacher*)

下面是刘慈欣同篇小说中对黄土高原上的一个村庄的描写,短短几句就描写出了一个荒凉、落寞的村庄,写作高度凝练。

例9 中文原文:

村庄静静地卧在月光下,像是百年前就没人似的。那些黄土高原上特有的平顶小屋,形状上同村子周围的黄土包没啥区别,在月夜中颜色也一样,整个村子仿佛已溶入这黄土坡之中。只有村前那棵老槐树很清楚,树上干枯枝杈间的几个老鸦窝更是黑黑的,像是滴在这暗银色画面上的几滴醒目的墨点。(刘慈欣,"乡村教师"《刘慈欣短篇小说合集》)

英文译文:

The village lay serenely in the moonlight, and it looked as if it had been abandoned for a hundred years. The small flat-roofed houses were almost indistinguishable from the mounds of soil surrounding them. In the muted colors of moonlight, it was as if the entire place had dissolved back into the hills. Only the old locust tree could be seen clearly, a few black crows' nests scattered among its withered branches, like stark drops of black ink on a silver page. (Lanphier Adam 译, *The Village Teacher*)

三、陌生化

航空航天文学文本还有一个重要特征是陌生化,主要体现在两个方面:一是所描述的场景或事件的陌生化;二是航空航天术语的陌生化。

由于航空航天所描述的场景或事件不是普通人所能体验的,是地球上未曾发生过的,所以对读者来说是陌生化的。这在科幻类航空航天文学中体现得尤为明显。例如,刘慈欣的《流浪地球》一开头就描述出了常人所难以想象的地球停止转动的场景。

例 10 中文原文:

我没见过黑夜,我没见过星星,我没见过春天、秋天和冬天。我出生在刹车时代结束的时候,那时地球刚刚停止转动。地球自转刹车用了四十二年,比联合政府的计划长了三年。妈妈给我讲过我们全家看最后一个日落的情景,太阳落得很慢,仿佛在地平线上停住了,用了三天三夜才落下去。当然,以后没有"天"也没有"夜"了,东半球在相当长的一段时间里(有十几年吧)将处于永远的黄昏中,因为太阳在地平线下并没落深,还在半边天上映出它的光芒。就在那次漫长的日落中,我出生了。①

英文译文:

I've never seen the night, nor seen a star; I've seen neither spring, nor fall, nor winter. I was born at the end of the Reining Age, just as the Earth's rotation was coming to a final halt. The Reining lasted for 42 years, three years longer than the Unity Government had projected. My mother once told me about the time our family witnessed the last sunset. The Sun had ever so slowly crept toward the horizon, almost as if it had stopped moving altogether. In the end, it took three days and three nights to finally set. Naturally, that was the end of all "days" and all "nights". The Eastern Hemisphere was shrouded in perpetual twilight for a long time then, perhaps for a dozen years or so —with the Sun hiding just beyond the horizon—its rays reflected by half of the sky. It was during that long sunset that I was born.②

---

① 刘慈欣. 流浪地球[M]. 武汉:长江文艺出版社,2008.
② Liu Cixin. The Wandering Earth[M]. Trans. by Holger N. Beijing:Beijing Guomi Digital Technology Co., Ltd, 2012.

下面是郝景芳所著《北京折叠》中对折叠城市三层空间的描述,也是普通人难以理解的陌生化情景。

例11 中文原文:

折叠城市分三层空间。大地的一面是第一空间,五百万人口,生存时间是从清晨六点到第二天清晨六点。空间休眠,大地翻转。翻转后的另一面是第二空间和第三空间。第二空间生活着两千五百万人口,从次日清晨六点到夜晚十点。第三空间生活着五千万人,从十点到清晨六点,然后回到第一空间。时间经过了精心规划和最优分配,小心翼翼隔离,五百万人享用二十四小时,七千五百万人享用另外二十四小时。①

英文译文:

The folding city was divided into three spaces. One side of the earth was First Space, population five million. Their allotted time lasted from six o'clock in the morning to six o'clock the next morning. Then the space went to sleep, and the earth flipped. The other side was shared by Second Space and Third Space. Twenty-five million people lived in Second Space, and their allotted time lasted from six o'clock on that second day to ten o'clock at night. Fifty million people lived in Third Space, allotted the time from ten o'clock at night to six o'clock in the morning, at which point First Space returned. Time had been carefully divided and parceled out to separate the populations: Five million enjoyed the use of twenty-four hours, and seventy-five million enjoyed the next twenty-four hours.②

"黑洞"也是我们所不常见、未体验过的场景。下文描述的是黑洞中的漩涡运动的情形。

例12 英文原文:

Space Whirl

Black holes can spin, just as the Earth spins. A spinning hole drags space around it into a vortex-type, whirling motion. Like the air in a tornado, space whirls fastest near the hole's center, and the whirl slows as one moves outward, away from the hole. Anything that falls toward the hole's horizon, gets dragged, by the whirl of space, into a whirling motion around and around the hole, like a straw caught and dragged by a tornado's wind. Near the horizon, there is no way whatsoever to protect oneself against this

---

① 郝景芳.北京折叠[M].南京:江苏凤凰文艺出版社,2016.
② Hao J F. Folding Beijing[M]. Trans. by Ken Liu. London: Head of Zeus, 2017.

whirling drag.[①]

中文译文：

无可抵抗的空间"漩涡"

黑洞可以自旋，就好像地球会自转一样。自旋的黑洞会拉动它周围的空间进行一种漩涡式的回旋运动。就好像飓风中的空气在靠近黑洞中心的地方时空间回旋会变得越来越快，而越向外远离黑洞的地方会回旋得越慢一样，任何落向黑洞视界的东西都会被回旋的空间拉，不停地绕着黑洞做回旋运动，如同被飓风捕获和拖拽着的一根稻草。在视界附近，这种拖拽将变得无法抗拒。[②]

航空航天文学文本陌生化的另一方面的体现是所运用的术语的陌生化。下面的这段描述中就含有不少的天文学和物理学方面的专业术语，非专业人士读起来较有陌生感。

例 13　中文原文：

太阳的灾变将炸毁和吞没太阳系所有适合居住的<u>类地行星</u>，并使所有<u>类木行星</u>完全改变形态和轨道。自第一次氦闪后，随着<u>重元素</u>在太阳中心的反复聚集，太阳氦闪将在一段时间反复发生，这"一段时间"是相对于恒星演化来说的，其长度可能相当于上千个人类历史。所以，人类在以后的太阳系中已无法生存下去，唯一的生路是向外太空恒星际移民，而照人类目前的技术力量，全人类移民唯一可行的目标是<u>半人马座比邻星</u>，这是距我们最近的恒星，有 4.3 <u>光年</u>的路程。以上看法人们已达成共识，争论的焦点在移民方式上。[③]

英文译文：

This stellar disaster would not only annihilate and consume every inhabitable <u>telluric planet</u> in the solar system, but it would also forever change the nature and orbits of the <u>Jovian planets</u>. After the primary <u>helium flash</u>, the <u>heavy elements</u> would re-accumulate in the core of the Sun and further helium flashes would repeatedly occur for a period of time. This was a "period" in the stellar sense, lasting many, many thousands of human life times. All of this made it impossible for humanity to continue living in the solar system, leaving only one last resort: The migration to another star. The technology of the time allowed for only one destination for this migration. That destination was <u>Proxima Centauri</u>, the star closest to ours, a mere 4.3 <u>light-years</u> away. But while it was easy to reach a consensus on the

---

① Kip T. The Science of Interstellar[M]. New York: W. W. Norton & Company, 2014.
② 基普·索恩. 星际穿越[M]. 苟利军,王岚,李然,等译. 杭州:浙江人民出版社,2015.
③ 刘慈欣. 流浪地球[M]. 武汉:长江文艺出版社,2008.

goal of the migration, the means were far more controversial.①

练 习 题

一、基础练习:请翻译以下句子

1. Inside our own galaxy, there are roughly 100 million smaller black holes: holes that typically are between about three and thirty times as heavy as the Sun.②

2. We know this not because we've seen evidence for all these, but because astronomers have made a census of heavy stars that will become black holes when they exhaust their nuclear fuel.③

3. Barbara put on a modest shirtdress, the hem skimming her calves.④

4. She buttoned the collar down conservatively but then, as a nod to her femininity, cinched the belt at her waist, showing off her slim figure.⑤

二、拓展练习:请翻译以下段落

1. Before the voting commenced, however, Barbara's co-workers hoisted her atop a convertible and drove around the lab. With the wind blowing through her hair, she smiled and waved. She felt a little silly and laughed nervously. Barbara might not be the prettiest girl in the lab, but she was sociable and easy to work with. All the computers were rooting for her. She imagined the director of the institute crowning her at the summer dance.⑥

2. Yet she didn't dwell on her possible victory, since it was merely a lighthearted affair. No one took it too seriously. Barbara, representing the computing section, was competing against Lois Labee, a chemist, and Margaret Anderson, who worked in the research design division. They were all young, beautiful, and very good at their demanding jobs. As odd as it seems by today's standards, the beauty contest was a result of JPL's progressive hiring practices. As

---

① Liu Cixin. The Wandering Earth[M]. Trans. by Holger N. Beijing: Beijing Guomi Digital Technology Co., Ltd, 2012.
② Kip T. The Science of Interstellar[M]. New York: W. W. Norton & Company, 2014.
③ 同②。
④ Holt N. Rise of the Rocket Girls[M]. New York: Little, Brown and Company, 2016.
⑤ 同④。
⑥ 同④。

the bouquets were handed out and an attractive woman crowned the winner, the competition was unintentionally highlighting the presence of educated young women working at JPL. After all, other laboratories would have found it impossible to hold such a contest in the 1950s; they simply didn't hire enough women.①

第三节 航空航天文学的翻译概况

翻译在大类上可分为文学翻译与非文学翻译。整体而言,非文学翻译是翻译的主体,文学翻译,即小说、诗歌、戏剧等的翻译,仅占翻译总量的百分之几。虽然所占比重较小,但文学翻译的意义重大,因为文学是一个民族思想和感情的重要载体,也代表着一个民族的艺术和智慧。特别是在当今"讲好中国故事、传播好中国声音"的时代背景下,中国文学在英语世界的翻译与传播,既具有重要的理论意义,也具有重要的实践意义。

邵璐②将中国当代文学的海外传播划分为了三个历史阶段,这三个历史阶段分别在译介选材、译介主体和译介规模三个维度呈现出不同的时代特征:(1)冷战背景下的中国当代文学海外翻译与传播(1949—1979年)。在冷战背景下,中国当代文学在英语世界的译介刚刚起步。这一阶段,当代文学海外传播的规模较小,且深受意识形态影响,在官方机构和官方资助译者群的支持下,无产阶级文学作品获得了较好的译介。(2)改革开放背景下的中国当代文学海外翻译与传播(1980—2001年)。1980—2001年,中国迎来改革开放大潮,政治、经济和文化发生了天翻地覆的变化。中国当代文学在这一背景下迅速发展,其海外传播也进入新时期。在译介选材方面,随着译者自主选择权的扩大,这一时期的选材内容相比于前一时期更加丰富。"改革"和"开放"主题更是成为向外译介的主旋律。在译介主体方面,国内翻译教育水平的提升促进了翻译力量的壮大,国外也涌现出葛浩文(Howard Goldblatt)、杜博妮(Bonnie S. McDougall)和蓝诗玲(Julia Lovell)等致力于翻译中国文学的汉学家,他们将中国文学作品介绍到海外。(3)全球化背景下的中国当代文学海外翻译与传播(2002年至今)。2002年至今,在全球化的推动下,中国文学翻译与传播迎来了繁荣期,选材上体现出多样性和经典性的特点,不少经典文学作品获得翻译或重译,科幻小说在被译介的同时呈现出影视化的趋势,网络小说也通过各类在线阅读平台进入了英语世界读者的视野。译介主体日益丰富,国家机构、海内外民间翻译力量和在数字媒体时代逐渐成熟的众包翻译团队共同发力。

---

① Holt N. Rise of the Rocket Girls[M]. New York: Little, Brown and Company, 2016.
② 邵璐.中国当代文学在英语世界的翻译与传播:框架、思路与方法[J].扬子江文学评论,2022(6):17-22.

就航空航天文学翻译而言,目前这方面的翻译还较少。在英译汉方面,航空航天文学翻译主要体现在两方面:一方面是描述战争的航空航天文学翻译,如江苏凤凰文艺出版社出版了张雅楠翻译的英国肯·福莱特的描述第二次世界大战的《大黄蜂奇航》(*Hornet Flight*),广西师范大学出版社出版了李平、徐菊清翻译的描述第二次世界大战的《飞行战犬》(*War Dog*),但数量不多;另一方面是科幻翻译,数量较大,如大连理工大学出版社出版了英国威尔斯科幻小说集,包括《月球上的第一批来客》(*The First Men in the Moon*)、《彗星来临》(*In the Days of the Comet*)、《大空战》(*The War in the Air*)、《新人来自火星》(*Star Begotten*),四川科学技术出版社出版了加拿大作家的"星辰舞"系列科幻作品,重庆出版社出版了美国科幻作家的《火星三部曲》(*Mars Trilogy*)系列。在汉译英方面,目前我国航空航天文学翻译主要集中在科幻翻译方面。21世纪以后,中国科幻文学的外译空前活跃,特别是《三体》(*The Three-body Problem*),通过美籍华人 Ken Liu(刘宇昆)的译介于 2015 年获得世界科幻文学的最高奖项——第 73 届"雨果奖"之最佳长篇小说奖,郝景芳的《北京折叠》(*Folding Beijing*)同样通过刘宇昆的译介于 2016 年获得第 74 届"雨果奖"之最佳中短篇小说奖,中国科幻文学对外译介也进入了一个空前繁荣的时期,中国科幻文学在海外拥有了较大的读者群,并产生了较好的传播效果[①]。根据高茜、王晓晖[②]的统计,2000—2020 年,中国科幻小说英译数量共计 214 种,其中长篇小说 10 部,中短篇小说 204 篇。这些外译的科幻小说中就有大量的航空航天类文学翻译,如刘慈欣的《三体》(*The Three-body Problem*)、《流浪地球》(*The Wandering Earth:Classic Science Fiction Collection*)、郝景芳的《北京折叠》(*Folding Beijing*)、刘宇昆主编的《看不见的星球》(*Invisible Planets:Contemporary Chinese Science Fiction in Translation*,2016)和《碎星星》(*Broken Stars:Contemporary Chinese Science Fiction in Translation*,2019)。对于 21 世纪中国科幻的英译热潮,高茜、王晓晖[③]认为有三方面的因素不容忽视:(1) 就中国科幻文学作品自身而言,其原作及译作质量上乘,这是成功进入英语世界的基础,也是中国科幻文学作品"走出去"务必坚持的根本所在;(2) 从目标语文学系统考量,主要是美国对非英语科幻文学作品的态度逐渐开放,而促成二者相遇的重要契机便是中外科幻界日趋频繁的交流;(3) 我国在国家层面积极支持科幻创作和中国文化"走出去",特别是在刘慈欣、郝景芳获奖后,科幻热潮不断发酵,中国本土出版社和科幻文化企业利用利好政策,积极拓展海外版图,探索出了有效的输出模式。

[①] 熊兵.中国科幻文学译介研究二十年(2000—2020):回顾、反思与展望[J].外国语文研究,2022(6):36-48.
[②] 高茜,王晓晖.中国科幻小说英译发展述评:2000—2020 年[J].中国翻译,2021(5):57-64.
[③] 同②。

练 习 题

一、基础练习：请翻译以下句子

1. The mission's goal, as defined by President Kennedy in 1961, was to "land a man on the moon and return him safely to Earth."①

2. Once it was circling the moon, a second spacecraft, the lunar module, would bring two astronauts down to the surface, leaving one man in the cone-shaped command module.②

3. This really superb piece of abstract thinking led Moon-Watcher, after only three or four minutes, to a deduction which he immediately put to the test.③

4. They could never guess that their minds were being probed, their bodies mapped, their reactions studied, their potentials evaluated.④

二、拓展练习：请翻译以下段落

It whirled around, throwing its insanely daring tormentor against the wall of the cave. Yet whatever it did, it could not escape the rain of blows, inflicted on it by crude weapons wielded by clumsy but powerful hands. Its snarls ran the gamut from pain to alarm, from alarm to outright terror. The implacable hunter was now the victim, and was desperately trying to retreat.⑤

三、阅读实践

1. 请阅读一部中文的航空航天文学作品。
2. 请阅读一部英文的航空航天文学作品。
3. 请阅读一部汉译英或英译汉的航空航天文学作品。

---

① Holt N. Rise of the Rocket Girls[M]. New York: Little, Brown and Company, 2016.
② 同①。
③ Clarke C A. 2001: A Space Odyssey[M]. UK: Pearson Education Ltd & Penguin Books Ltd, 1968.
④ 同③。
⑤ 同③。

第二章　航空航天文学翻译原则

在我国，翻译已有两千多年的历史。在古代，周代就有了关于翻译活动的明确记载。《礼记·王制》中记载："中国戎夷，五方之民，皆有其性也，不可推移。东方曰夷，被发文身，有不火食者矣。南方曰蛮，雕题交趾，有不火食者矣。西方曰戎，被发衣皮，有不粒食者矣。北方曰狄，衣羽毛穴居，有不粒食者矣。中国、夷、蛮、戎、狄，皆有安居、和味、宜服、利用、备器。五方之民，言语不通，嗜欲不同。达其志，通其欲，东方曰寄，南方曰象，西方曰狄鞮，北方曰译。"

唐代孔颖达在《礼记正义》中一一作了解释："其通传东方之语官曰寄，言传寄外内语言。通传南方语官谓之曰象者，言放象外内之言。其通传西方语官谓之狄鞮者，鞮，知也，谓通传夷狄之语，与中国相知。其通传北方语官曰译者，译，陈也，谓陈说外内之言。"

虽然翻译已有悠久的历史，但翻译中的问题和困难似乎始终没有得到很好的解决。翻译不易，自古至今，但凡从事过翻译的人都有此体会。我国译论的开篇——支谦的"法句经序"对翻译发出的感叹就是"传实不易"："夫诸经为法言，《法句》者，犹法言也。近世葛氏传七百偈，偈义致深，译人出之，颇使其浑漫。惟佛难值，其文难闻，又诸佛兴，皆在天竺。天竺言语与汉异音，云其书为天书，语为天语，名物不同，传实不易。"

近代著名翻译家严复对翻译发出的感叹也是"译事三难"："译事三难：信、达、雅。求其信已大难矣，顾信矣不达，虽译犹不译也，则达尚焉。"

我国著名学者钱钟书认为，翻译一部文学作品要比研究一部文学作品难得多："我们研究一部文学作品，事实上往往不能够而且不需要一字一句都透彻了解的。有些字、词、句以至无关紧要的章节都可以不求甚解；我们一样写得出头头是道的论文，完全不必声明对某字、某句和某节缺乏了解，以表示自己特别诚实。翻译可就不同。原作里没有一个字可以滑溜过去，没有一处困难躲闪得了。"[1]

翻译困难，文学翻译尤其困难。因此，为了解决翻译中的各种问题和困难，翻译研究者们提出了诸多的翻译原则。

---

[1] 钱钟书. 林纾的翻译[M]//翻译论集. 北京：商务印书馆，1984.

第一节　文学翻译原则

提及翻译原则,我们首先想到的就是泰特勒的翻译三原则。泰特勒(Alexander Fraser Tytler,1747—1814年)是18世纪英国著名翻译理论家。泰特勒在《论翻译的原则》(*Essay on the Principles of Translation*,1790)一书中提出了"翻译三原则",为我国译学界所熟知。这三条原则是:

(1) 译作应该完全复写出原作的思想。
(2) 译作的风格和手法应与原作属于同一性质。
(3) 译作应具备原作所具有的通顺。

泰特勒的英文原文为:

(1) The translation should give a complete transcript of the idea of the original work.

(2) The style and manner of writing should be of the same character with that of the original.

(3) The translation should have all the ease of the original composition.

虽然泰特勒提出了这三条翻译原则,但他也深知这三个目标是很难达到的。泰特勒对第三条原则进行了解释,他把翻译比作画家临摹,但比临摹更难:

To one who walks in trammels, it is not easy to exhibit an air of grace and freedom. It is difficult, even for a capital painter, to preserve in a copy of a picture all the ease and spirit of the original; yet the painter employs precisely the same colours, and has no other care than faithfully to imitate the touch and manner of the picture that is before him. If the original is easy and graceful, the copy will have the same qualities, in proportion as the imitation is just and perfect. The translator's task is very different: He uses not the same colours with the original, but is required to give his picture the same force and effect. He is not allowed to copy the touches of the original, yet is required, by touches of his own, to produce a perfect resemblance. The more he studies a scrupulous imitation, the less his copy will reflect the ease and spirit of the original. How then shall a translator accomplish this difficult union of ease with fidelity? To use a bold expression, he must adopt the very soul of his author, which must speak through his own organs.

(一个人如果走路受到束缚,要他走得姿势优雅自在是不容易的。即使是一个优秀的画家,临摹一幅作品要保留原作的流畅和精神,就更难了。可是,画家用的是完全相同的颜色,他只要忠实地模仿在他面前的原作的笔法和表现手法就可以了。如果原作流畅优美,只要摹本完美无缺,就可具有原作的一切特征。然而,译者的工作完全不同。他用的不是与原作相同的颜色,但必须使自己的"画"具有同样的感染力和效果。但他不能照搬原作的笔法,而必须用自己的笔法来译出完美的"摹本"。他对摹本的研究越一丝不苟,他的译作离原作的通顺和精神就越远。那么,译者怎样才能做到既忠实又通顺呢?说句大胆的话,译者必须用自己的话表达原作者的精神。)①

当然,提到翻译的原则,也不能不提及严复的"信、达、雅"三原则。1898年,严复在《天演论·译例言》开篇中提出"信、达、雅"三原则。严复不仅提出了"信、达、雅"的一系列的基本概念,还较为科学地阐述了这三个概念间的逻辑体系:"求其信已大难矣";"信矣不达,虽译犹不译";"信达而外,求其尔雅"。刘宓庆②认为,"信达雅"是三环联袂的统一体:要做到"信",就必须"达";要做到"达",途径是"雅"。严复的"信、达、雅"翻译三原则可以称得上是中国近现代第一个体系化的翻译理论。对于"信、达、雅"学说为何成了中国译学颠扑不破的"真理",武光军、蒋雨衡③认为,主要有三个方面的原因:第一,该学说与各国翻译理论具有共性,符合普遍的翻译标准。这里主要指的是"信"和"达"的提出。"信"和"达"应该说是各国普遍的翻译标准。第二,该学说具有中国特色,即具有个性。这里主要指的是"雅"的提出。这是中国翻译理论的特色成分。"信、达、雅"的特征在"雅"。第三,"信、达、雅"学说深植于中国的传统哲学和文章学,具有深厚的传统根基,同时其命题方式也符合中国传统哲学——美学的命题方式。严复把译事之难概括为"信、达、雅",既符合翻译实际,又与中国传统哲学——美学思想及命题相承袭。这也是"三难之说"历久不衰的原因之一。中国传统哲学——美学讲求以三个命题相济相容,称为"三表法"。

但文学翻译大大不同于其他文体的翻译。在实际的文学翻译中,文学翻译对"信"的要求往往不高,对准确性的要求也低。例如,葛浩文译的姜戎的《狼图腾》不仅有对典故引用和情节方面的大幅度的删减,在具体的句子层面的翻译中也进行了修改,例如:

例1 中文原文:
陈阵急忙下马,铲清扫净了雪。老人蹲下身,用小铁镐在冻得不太深的土地上刨出一个直径约40厘米,深约15厘米的圆坑,坑中还有一个小坑。(姜戎,《狼图腾》)

英文译文:
Chen dismounted and started digging in the snow, while Bilgee crouched down and scraped out <u>a circle about a foot and a half across and a couple of inches deep</u> with his little spade.(葛浩文译,*Wolf Totem*)

我们看到,姜戎的《狼图腾》中较为准确的"一个直径约40厘米,深约15厘米的

---

① 郭建中.泰特勒翻译三原则中译辨正[J].中国翻译,2013(3):68-70.
② 刘宓庆.翻译美学导论[M].北京:中国对外翻译出版公司,2019:60-61.
③ 武光军,蒋雨衡.严复"信、达、雅"来源考辨及其译学意义重释[J].中国翻译,2021(3):50-56.

圆坑"并没有在葛浩文的译文中被忠实地翻译。在其他翻译家翻译的作品中,这种不"信"的现象也比比皆是,例如沙博理译的高云览的《小城春秋》也有不少的省译。

例 2 中文原文:

原来古冢室不过是一间装置各种古董字画的暗室。刘眉把一把一百烛光的电灯扭亮,热心地指着那些历代的铜钱、陶舰、人头骨、贝、蚌、雕花的木器、甲骨和一些擦得发亮的外国瓷器、杯盘,叫客人们观赏。可惜客人们缺乏观赏家的兴致,只走马观花地过一下眼,就走出来了。①

英文译文:

Liu Mei turned on a hundred-watt bulb and proudly exhibited his antiques and curios. His guests displayed only a desultory interest.②

我国著名翻译家傅雷是法语文学翻译的大家,对于文学翻译的原则和境界,傅雷提出了著名的"神似论"。傅雷认为,在文学翻译中,我们追求的应是"神似",而不是"形似":以效果而论,翻译应当像临画一样,所求的不在形似而在神似。以实际工作论,翻译比临画更难。临画与原画,素材相同(颜色、画布、纸或绢),法则相同(色彩学、解剖学、透视学)。译本与原作,文字既不侔,规则又大异。各种文字各有特色,各有无可模仿的优点,各有无法补救的缺陷,同时又各有不能侵犯的戒律。像英、法,英、德那样接近的语言,尚且有许多难以互译的地方;中西文字的扞格远过于此,要求传神达意,铢两悉称,自非死抓字典,按照原文句法拼凑堆砌所能济事。(傅雷《高老头》重译本序",1951)

我国著名学者钱钟书先生③在《林纾的翻译》一文中讨论"译"的来源时提出:翻译的最高境界是"化",即"化境"的翻译原则:汉代文字学者许慎有一节关于翻译的训诂,意蕴颇为丰富。《说文解字》卷六《口》部第二十六字:"囮,译也。从'口','化'声。率鸟者系生鸟以来之,名曰'囮',读若'讹'。"南唐以来,"小学"家都申说"译"就是"传四夷及鸟兽之语",好比"鸟媒"对"禽鸟"所施的引"诱","讹"、"讹"、"化"和"囮"是同一个字。"译"、"诱"、"媒"、"讹"和"化"这些一脉通连、彼此呼应的意义,组成了研究诗歌语言的人所谓"虚涵数意"(manifold meaning),把翻译能起的作用("诱")、难于避免的毛病("讹")、所向往的最高境界("化"),仿佛一一透示出来了。

但钱钟书认为翻译所起的作用为"诱",也就是说,翻译在文化交流中起着一种"媒婆"的作用,是两种文化间的"沟通""桥梁",目的是要引导人向"原作过渡",不仅是"要省人家的事,免得他们去学外文、读原作"。这很不同于泰特勒所提出的译文应该复制出原文的内容和风格的翻译原则及翻译主张。这主要是因为,在钱钟书④看

---

① 高云览. 小城春秋[M]. 北京:人民文学出版社,2007.
② Gao Y L. Annals of a Provincial Town[M]. Trans. by Shapiro S. Beijing: Foreign Languages Press, 1959.
③ 钱钟书. 林纾的翻译[M]//翻译论集. 北京:商务印书馆,1984.
④ 同①.

来，翻译活动存在难以避免"讹"：一国文字和另一国文字之间必然有距离，译者的理解和文风跟原作品的内容和形式之间也不会没有距离，而且译者的体会和他自己的表达能力之间还时常有距离。从一种文字出发，积寸累尺地度越那许多距离，安稳到达另一种文字里，这是很艰辛的历程。一路上颠顿风尘，遭遇风险，不免有所遗失或受些损伤。因此，译文总有失真和走样的地方，在意义或口吻上违背或不尽贴合原文。那就是"讹"，西洋谚语所谓"翻译者即反逆者"。

鉴于文学翻译中传统的"忠实"的翻译原则与翻译实践中存在的难以避免的"讹"之间的矛盾，"创造性叛逆"这一概念在文学翻译中的产生就是自然而然的事了。"创造性叛逆"最初由法国文艺社会学家罗伯特·埃斯卡皮（Robert Escarpit）提出：翻译总是一种创造性叛逆。说翻译是叛逆，那是因为它把作品置于一个完全没有预料到的参照体系（指语言）里；说翻译是创造性的，那是因为它赋予作品一个崭新的面貌，使之能与更广泛的读者进行崭新的文学交流，还因为它不仅延长了作品的生命，而且又赋予它第二次生命。[1]

在我国，较早大力推动和提倡在文学翻译中运用创造性叛逆的学者是谢天振。在其代表作《译介学》中，谢天振认为，不管译者对作者持什么态度，从译介学的观点看，其中的创造性叛逆都是不可避免的。谢天振把创造性界定为"译者以自己的艺术创造才能去接近和再现原作的一种主观努力"，把叛逆定义为"在翻译过程中译者为了达到某一主观愿望而造成的一种译作对原作的客观背离"[2]。此外，谢天振还认为创造性与叛逆性是一个和谐的有机体：在文学翻译里，无论是译作胜过原作，还是译作不如原作，这些现象都是由文学翻译的创造性与叛逆性所决定了的。如果说，文学翻译中的创造性表明了译者以自己的艺术创造才能去接近和再现原作的一种主观努力，那么文学翻译中的叛逆性，就是反映了在翻译过程中译者为了达到某一主观愿望而造成的一种译作对原作的背离。但是，这仅仅是从理论上而言，在实际的文学翻译中，创造性与叛逆性其实是根本无法分割的，它们是一个和谐的有机体。[3]

"创造性叛逆"这一概念引入我国文学翻译界后，引起了翻译研究学者的广泛关注。吴雨泽甚至认为，创造性叛逆不啻成为解释所有译作与原作差异现象的万能良药：长期以来，"忠实"一直被视为亘古不变的翻译标准，这一点可谓中外皆然。从"按本"到"求信"，再到"神似"与"化境"，凡此等等，尽管措辞不一，表述有别，言者所重不过是"忠实"之于翻译不可或缺的本体或本质属性。的确，作为一种独特的认知活动，翻译终究无法摆脱原文的束缚，缺少了"忠实"这一标准的制约，也就失去了衡量与评价译本的客观尺度。然而自 20 世纪后半叶以来，随着翻译研究"文化转向"理念的输入，描写译学受到了国内译界中人的广泛青睐，与此同时，在创造性叛逆

---

[1] 罗贝尔·埃斯卡皮. 文学社会学[M]. 王美华,于沛,译. 合肥:安徽文艺出版社,1987.
[2] 谢天振. 译介学[M]. 上海：上海外语教育出版社,1999.
[3] 同[2].

(creative treason)、误读(misreading)等观点的影响下,"忠实"标准也逐渐失去了对翻译操作的制约作用,或至少受到了很大程度的冲击,创造性叛逆不啻成为解释所有译作与原作差异现象的万能良药。①

罗新璋也是大力提倡发挥译者创造性和彰显译者主体性的翻译家,他认为:作家运思命笔,自应充分发挥主体的创造力量,译者在翻译时难道就不需要扬起创造的风帆?须知译本的优劣,关键在于译者,在于译者的译才,在于译者的译才是否得到充分施展。重在传神,则要求译者能入乎其内,出乎其外,神明英发,达意尽韵。翻译理论中,抹杀译者主体性论调应少唱,倒不妨多多研究如何拓展译者的创造天地,于局限中掌握自由。②

董明③举了郭沫若译苏格兰诗人彭斯的名诗"A Red, Red Rose"的例子,说明译者与诗人的水乳交融和译者的创造性的体现:

例3 英文原文:
O My Luve's like a red, red rose,
That's newly sprung in June;
O My Luve's like the melody,
That's sweetly played in tune.
(Robert Burns)

中文译文:
吾爱吾爱玫瑰红,
六月初开韵晓风;
吾爱吾爱如管弦,
其声悠扬而玲珑。
(郭沫若译)

董明④认为,无论是从内容还是从形式上看,郭沫若的译文都不能说与原诗绝对对等,如"吾爱像红玫瑰"译成了"吾爱吾爱玫瑰红";原诗音步四步与三步、二行与四行押韵,译文变成了古汉语七绝形式;为了保持诗行的平衡,译文两次重复"吾爱",但这种创造性翻译却十分真实地再现了诗人真挚的情感。

的确,创造性叛逆可在一定程度上解释译文与原文间存在的差异。但我们不应忘却的是,无论是对"创造性"还是"叛逆",目前都没有较为准确的界定,自然,对两者结合后的"创造性叛逆"目前也没有较为准确的界定。原文与译文在多大程度上的偏离是创造性?原文与译文在多大程度上的偏离是叛逆?目前,我们尚无法给出答案。

---

① 吴雨泽. 在"忠实"标准的观照下:重释文学翻译中的创造性叛逆[J]. 西安外国语大学学报, 2013(4): 123-126.
② 罗新璋. 中外翻译观之"似"与"等"[J]. 世界文学, 1990(2): 285-286.
③ 董明. 文学翻译中的创造性叛逆[J]. 外语与外语教学, 2003(8): 46-49.
④ 同③。

许钧就提出了对这方面的忠告:翻译的创造性何在,如何认识翻译的创造性,在今天看来,在理论上存在着一定程度的分歧。由于思想认识上的不一致和理论上的分歧,在实践上必然会造成混乱。确实,我们注意到,在今天的译坛,有的译者以"创造"为名,行"背叛"之实,翻译时不细读原作,不顾及原作的底蕴与风格,随心所欲地加以处理。①

练 习 题

一、基础练习:请翻译以下句子

1. "A scandal if that land is taken for building!" he said, chopping up a piece of chocolate cake viciously. "More larks there to the square yard than anywhere else in England!"②

2. "And if the old Tartar finds out, it is all one to me!" he added sturdily, tucking it behind the sack which shrouded it.③

3. "Then I'll come definitely every Orchestra night," I promised. "I should have thought of it before. It's the least a godmother can do."④

4. Hundred Acre Field and its spacious neighbors were among the more fruitful parts of Mr. Miller's farm.⑤

二、拓展练习:请翻译以下段落

黄昏以前老道士用红绿纸剪了一些花朵,用黄泥做了一些烛台。天断黑后,棺木前小桌上点起黄色九品蜡,燃了香,棺木周围也点了小蜡烛,老道士披上那件蓝麻布道袍,开始了丧事中的绕棺仪式。老道士在前面拿着小小纸幡引路,孝子第二,马兵殿后,绕着那具寂寞棺木慢慢转着圈子。⑥

第二节 航空航天文学翻译:英译汉的原则

在上一节中,我们分析了文学翻译的总原则。当然,各类文学翻译都有共同之

---

① 许钧."创造性叛逆"与翻译主体性的确立[J].中国翻译,2003(1):6-11.
② 宋楠.语境理论背景下文学翻译策略的选择和应用:以英语小说《乡村风暴》翻译实践为例[J].牡丹江大学学报,2015,24(5):135-138+147.
③ 同②。
④ 同②。
⑤ 同②。
⑥ 沈从文.沈从文全集:第8卷[M].太原:北岳文艺出版社,2002.

处,但翻译是有方向性的差异的,例如,英译汉和汉译英应该有不同的翻译原则。这主要是因为两者的受众不同,前者的受众是以中文为母语者,后者的受众是以英语为母语者,不同的受众自然对翻译会有不同的需求。

对于航空航天文学的英译汉翻译原则,我们认为主要有两个方面:一是在语言上:既求达,也求雅。这是由中文的文体特点决定的,"言之无文,行之不远"。二是在内容上:求神似,不求形似。这意味着译文可根据目的语需求对原文的内容和语言的形式进行适度的修改或增、省等翻译操作。下面,我们通过航空航天文学文本中的景物描写、人物描写及事件描写的翻译具体实例来对这两方面的翻译原则进行探讨。

一、景物描写的翻译

例4 英文原文:

Let me describe this room to you in detail. It was perhaps eight feet by seven in area and rather higher than either of these dimensions; the ceiling was of plaster, cracked and bulging in places, gray with the soot of the lamp, and in one place discolored by a system of yellow and olive-green stains caused by the percolation of damp from above. The walls were covered with dun-colored paper, upon which had been printed in oblique reiteration a crimson shape, something of the nature of a curly ostrich feather, or an acanthus flower, that had in its less faded moments a sort of dingy gaiety. There were several big plaster-rimmed wounds in this, caused by Parload's ineffectual attempts to get nails into the wall, whereby there might hang pictures. One nail had hit between two bricks and got home, and from this depended, sustained a little insecurely by frayed and knotted blind-cord, Parload's hanging bookshelves, planks painted over with a treacly blue enamel and further decorated by a fringe of pinked American cloth insecurely fixed by tacks. Below this was a little table that behaved with a mulish vindictiveness to any knee that was thrust beneath it suddenly; it was covered with a cloth whose pattern of red and black had been rendered less monotonous by the accidents of Parload's versatile ink bottle, and on it, leitmotif of the whole, stood and stank the lamp. This lamp, you must understand, was of some whitish translucent substance that was neither china nor glass, it had a shade of the same substance, a shade that did not protect the eyes of a reader in any measure, and it seemed admirably adapted to bring into pitiless prominence the fact that, after the lamp's trimming, dust and paraffin had been smeared over its exterior with a reckless generosity.①

中文译文：

让我详细描述一下这间屋子吧。房间大约有八英尺长，七英尺宽，高度倒比这两个数字大出不少。石膏做的天花板已有裂缝，到处凹凸不平，被煤油灯的烟熏得灰蒙蒙的，有一处被天花板上渗出来的潮气染成黄色和橄榄绿色。墙上贴着暗褐色的墙纸，上面斜印着暗红色的图案，有点像鸵鸟的卷毛，或是像已经黯然失色、行将枯萎的茛苕花。墙上有几块石膏糊的补疤，那是因为帕洛德想把钉子打进墙里却又徒劳无功而留下的痕迹，否则他会在那上面挂上几幅画呢。有一颗钉子在两块砖之间找到了立足之地，上面挂着帕洛德悬挂式书架，靠着磨损打结德窗帘绳勉强没有掉落，书架的板条上刷了一层油腻腻的蓝色磁漆，上面又用大头针粗粗钉了一缕美国油布作装饰。书架下方是一张小桌，似乎能把任何突然塞进来的双膝狠狠卡断；桌上蒙着一块桌布，帕洛德的多用墨水瓶不留神被打翻过，使桌布红黑相间的图案不再那么单调了，桌上醒目的地方立着一盏台灯，散发出燃料的气味。这盏灯是用发白的半透明的东西做成的，可那东西既非陶瓷又非玻璃。灯罩是用相同的材料制作的，一点儿也起不到保护眼睛的作用，它似乎极能适应使下述事实更加显著的需求。那就是，此灯一经点燃，灰尘和煤油就会满不在乎、随随便便地在灯罩外面安营扎寨。②

这一段是英国科幻作家威尔斯在其科幻小说《彗星来临》中对一个破旧小屋的描写，可谓是入木三分。威尔斯先描写了这间破旧小屋中的大小，然后自上而下地描写了天花板、墙壁和挂在墙上的书架，最后描写了小屋地面上的小桌和台灯。再加上，威尔斯对这些物件的具体的描述：灰蒙蒙的天花板、破败的墙壁、几乎要掉落的书架、恐怖的小桌、难闻的油灯、不伦不类灯罩，读后这间小屋的丑陋和压迫感可谓是立刻令人生厌，不愿踏进半步。威尔斯的原作描写得精彩，穆雷的翻译同样也翻译得精彩。

从内容上来讲，中文的译文可谓是达到了"神似"：读完中文译文，这间小屋的破败感也同样跃然纸上，栩栩如生。特别是在较难翻译的地方，译者都展现出了自己的创造性的水平。例如，该段的最后一句（after the lamp's trimming, dust and paraffin had been smeared over its exterior with a reckless generosity.）是较难翻译的，特别是其中的"reckless generosity"这一用法更难翻译：一是因为这两个词本来是用来描述人的，不是描述物的；二是这种搭配是不好理解的。译者将此句译为"那就是，此灯一经点燃，灰尘和煤油就会满不在乎、随随便便地在灯罩外面安营扎寨。"可谓是译笔生花，神来之笔。将"smeared"译为"安营扎寨"，似是"误译"，实是"神译"，带出了"reckless generosity"这两词拟人的底蕴，不是为了译"smeared"而译"smeared"，而是瞻前顾后，弃形译神。对于"reckless generosity"这一令人费解的，似乎难以直译

---

① Herbert G W. In the Days of the Comet[M]. Auckland: The Floating Press, 2009.
② 赫伯特·乔治·威尔斯. 彗星来临[M]. 穆雷, 译. 大连：大连理工大学出版社, 2018.

的英语搭配形式,译者将其一分为二,译为"满不在乎、随随便便地",同样可谓是神来之笔,既没有过度翻译,也没有欠额翻译,恰到好处。

从语言上来讲,中文译文不仅通顺流畅,而且还用词用语严谨、生动、深刻,可谓是既达又雅。对于原文的第二句(It was perhaps eight feet by seven in area and rather higher than either of these dimensions; the ceiling was of plaster, cracked and bulging in places, gray with the soot of the lamp, and in one place discolored by a system of yellow and olive-green stains caused by the percolation of damp from above.),译文(房间大约有八英尺长,七英尺宽,高度倒比这两个数字大出不少。石膏做的天花板已有裂缝,到处凹凸不平,被煤油灯的烟熏得灰蒙蒙的,有一处被天花板上渗出来的潮气染成黄色和橄榄绿色。)可谓是明白晓畅。译者根据中文的特点进行了多处增添。"eight feet by seven in area"译为了"八英尺长,七英尺宽",把"长"和"宽"两个维度增添出来,在中文中更易懂。"the ceiling was of plaster, cracked and bulging in places, gray with the soot of the lamp, and in one place discolored by a system of yellow and olive-green stains caused by the percolation of damp from above."这部分在英语原文中为静态性的描述,译文将过程性的动作(做的、熏得、渗出来)添加出来,符合了汉语的动态性特征。特别是"caused"没有被直译为"使得或导致",而是根据上下文译为了"染成",恰到好处。原文的第三句(The walls were covered with dun-colored paper, upon which had been printed in oblique reiteration a crimson shape, something of the nature of a curly ostrich feather, or an acanthus flower, that had in its less faded moments a sort of dingy gaiety.)是含有一个较长的定语从句的长句,在结构上是较难翻译的。译者将其译为了四个短句:"墙上贴着暗褐色的墙纸,上面斜印着暗红色的图案,有点像鸵鸟的卷毛,或是像已经黯然失色、行将枯萎的莨苕花。"通顺流畅。"printed in oblique reiteration"译为"斜印着",简洁准确,又达又雅。"something of the nature of a curly ostrich feather"省去"the nature of",译为"有点像鸵鸟的卷毛",符合中文具象化的表达特征,通顺自然。

当然,译文中也尚有可商榷之处。该段最后一句话的"it seemed admirably adapted to bring into pitiless prominence the fact that, after the lamp's trimming, dust and paraffin had been smeared over its exterior with a reckless generosity."这部分,译为了"它似乎极能适应使下述事实更加显著的需求。那就是,此灯一经点燃,灰尘和煤油就会满不在乎、随随便便地在灯罩外面安营扎寨。"其中"它似乎极能适应使下述事实更加显著的需求。"译得较为生硬,受了原文字面和结构的束缚,在语言上似不达也不雅。

二、人物描写的翻译

例 5 英文原文:

But her bright brown hair, which had once flowed down her back in a jolly pig-tail tied with a bit of scarlet ribbon, was now caught up into an intricacy of pretty curves above her little ear and cheek, and the soft long lines of her neck; her white dress had descended to her feet; her slender waist, which had once been a mere geographical expression, an imaginary line like the equator, was now a thing of flexible beauty. A year ago she had been a pretty girl's face sticking out from a little unimportant frock that was carried upon an extremely active and efficient pair of brown-stockinged legs. Now there was coming a strange new body that flowed beneath her clothes with a sinuous insistence. Every movement, and particularly the novel droop of her hand and arm to the unaccustomed skirts she gathered about her, and a graceful forward inclination that had come to her, called softly to my eyes. A very fine scarf—I suppose you would call it a scarf— of green gossamer, that some new wakened instinct had told her to fling about her shoulders, clung now closely to the young undulations of her body, and now streamed fluttering out for a moment in a breath of wind, and like some shy independent tentacle with a secret to impart, came into momentary contact with my arm.[1]

中文译文：

可是她那亮丽的褐发曾经垂在后背，用一根鲜红色的缎带扎成一束活泼的马尾巴，现在却在她小巧的耳朵和脸蛋上方烫成漂亮的大波浪。还有她那线条修长的脖颈，白色的衣裙拖到脚面。她苗条的腰身过去不过仅仅表现出身段而已，就像赤道一样是一条想象中的地理分界线，如今却变得柔软灵活，美不胜收。一年前，她曾拥有一张青春少女的脸庞，从普普通通的上衣中探出来，一双长腿充满活力，穿着棕色的长袜。如今，一副陌生的、崭新的身材就在眼前，曲线毕露，妙不可言，一举手，一投足，尤其是她用双臂拢住新颖别致的裙裾时，优雅大方，看得我眼都直了。一条精美绝伦的绿纱巾——我想你会管这玩意叫纱巾，某种刚刚觉醒的本能使她低头扫视了一下自己的肩头——紧裹着她曲线毕露、青春焕发的身躯，在一阵微风中飘动，宛如一些羞怯而无拘无束的触须要透露什么秘密，不时地碰着我的胳膊。[2]

在《彗星来临》中译本的前言中，《科幻世界》的陈俊对威尔斯的这部作品进行了解读：《彗星来临》以一种散文式的记述，缓步推进着一个看似俗套的三角恋故事，却在关键节点上通过一条漫不经心的暗线将一次情杀危机反转。故事的背景是一颗彗星即将接近地球的消息不断在剧情中跟进，而情节上讲述的则是一位四处碰壁的主人公，在接连的失败与对刚刚分手不久的女友立即寻获归宿的妒意之下，决定谋杀前

---

[1] Herbert G W. In the Days of the Comet[M]. Auckland: The Floating Press, 2009.
[2] 赫伯特·乔治·威尔斯. 彗星来临[M]. 穆雷，译. 大连：大连理工大学出版社，2018.

女友及其情夫。但就在下手的当晚,一颗彗星的尾巴扫过地球,通过与空气中的氮气反应产生了"绿色烟雾"给予世间人心以光明、友爱等善良的品质。于是世界变成了乌托邦,故事变成了大团圆。在《彗星来临》中,威尔斯想表达的主旨就是探索一种破除道德束缚、打破阶级壁垒的美好的乌托邦社会,在小说中最重要的体现就是威尔斯对男女主人公在彗星来临后消除私欲的过程的描述。

  上面这段描述的就是女主人公内蒂的形象。威尔斯的英语文学描写可谓是栩栩如生,跃然纸上。同样,中文译文可谓是相得益彰,毫不逊色,不仅传神,而且达雅。在多处难译之处,译者都展现出了神来之笔。第一句"But her bright brown hair, which had once flowed down her back in a jolly pig-tail tied with a bit of scarlet ribbon, was now caught up into an intricacy of pretty curves above her little ear and cheek"中,"was now caught up into an intricacy of pretty curves"是无法用中文直接对应转换的,较难翻译,译者用一个英语原文中并没有出现的"烫"字巧妙地化解了这一翻译难题。第二句"A year ago she had been a pretty girl's face sticking out from a little unimportant frock that was carried upon an extremely active and efficient pair of brown-stockinged legs."中的"sticking out"也是难以找到中文对应词的。译者透悟上下文语境,将该词译为"探出来",惟妙惟肖,可谓传神达意,不求形似,但求神似。第三句"Now there was coming a strange new body that flowed beneath her clothes with a sinuous insistence."的英文表达较为抽象,特别是"sinuous insistence"更是让人丈二和尚摸不着头脑,翻译起来似乎无从下手。译者舍弃形似,追求神似。整句话译为"如今,一副陌生的、崭新的身材就在眼前,曲线毕露,妙不可言",可谓透彻领悟,属"化境"之作。

  在语言方面,译文既达又雅。在达的方面,译者灵活地处理了原文中的长句。第一句"But her bright brown hair, which had once flowed down her back in a jolly pig-tail tied with a bit of scarlet ribbon, was now caught up into an intricacy of pretty curves above her little ear and cheek, and the soft long lines of her neck; her white dress had descended to her feet; her slender waist, which had once been a mere geographical expression, an imaginary line like the equator, was now a thing of flexible beauty."是一个非常长的句子,含有两个分号和多个定语从句。译者将该句首先化解成了中文译文中的三句,每句中又破除了长定语的困扰,译为"可是她那亮丽的褐发曾经垂在后背,用一根鲜红色的缎带扎成一束活泼的马尾巴,现在却在她小巧的耳朵和脸蛋上方烫成漂亮的大波浪。还有她那线条修长的脖颈,白色的衣裙拖到脚面。她苗条的腰身过去不过仅仅表现出身段而已,就像赤道一样是一条想象中的地理分界线,如今却变得柔软灵活,美不胜收。"这样的中文译文读来像是娓娓道来,毫无结构臃肿、生涩之感,通达流畅。最后一句的后半部分"and like some shy independent tentacle with a secret to impart, came into momentary contact with my arm"是作者威尔斯构思出来的一个极富想象力的比喻,给人以无限遐想的空间。译者完整再现了威尔斯的这一独创的比喻,译为了"宛如一些羞怯而无拘无束的触须要透露什么秘密,不时地碰着我的胳膊",读来晓畅自然,无异域生涩之感。译者用词典雅。第四句"Every movement, and particularly the novel droop of her hand and arm to the unaccustomed skirts she gathered about her, and a graceful forward inclination that had come to her, called softly to my eyes."译为"如今,一副陌生的、崭新的身材就在眼前,曲线毕露,妙不可言,一举手,一投足,尤其是她用双臂拢住新颖别致的裙裾时,优雅大方,看得我眼都直了。"可谓用词考究,如"举手""投足""拢""裙裾"等。当然,译文中也还有可商榷之处。例如,"一条精美绝伦的绿纱巾——我想你会管这玩意叫纱巾"中的"玩意"似不符合此处上下文之文意。

三、故事情节的翻译

例6 英文原文:

Their controlled descent through the mist was over in a matter of seconds. In spite of Pierre's desperate manoeuvres the German gunners quickly found their mark. Rounds ripped through the thin fuselage and shattered the Perspex cockpit. As smoke and fire bloomed from the port engine, Robert sensed that they were going down. They were barely two hundred feet above the snowbound earth when he saw the port propeller die completely, and felt the enemy fire tearing into their starboard engine.

Robert braced himself for the impact of a crash-landing or worse. The hard, frozen ground was rushing up to meet them, a wide expanse of glistening snow lit here and there a fiery red by the tracer fire. Barely minutes after they'd first been hit, the belly of the aircraft impacted with a terrible tearing of metal. The stricken warplane lifted once, settled again with an ear-piercing screech and ploughed towards a patch of dark woodland.

The doomed aircraft was thrown savagely around as its left flank caught on a thick trunk, and with a tearing of steel, the wing was ripped clean away. By the time it came to a juddering halt, half buried in the snow and with its crumpled nose cone embedded in the thick foliage, Robert had lost consciousness.[①]

中文译文:

仅仅几秒钟后,他们便无法在雾中控制飞机的下降幅度。尽管皮尔极力地变换飞行轨迹,他们还是被德国枪炮手发现了。一阵射击后,单薄的机身和驾驶舱玻璃出

---

① Lewis D. War Dog[M]. London: Sphere, 2013.

现了裂口，浓烟和火焰不断从左翼引擎中冒出，罗伯特感觉他们在向下坠落。他还注意到，飞机的左翼螺旋桨已完全失灵，敌人正猛攻他们的右翼引擎。此时，他们距白雪皑皑的大地仅有200英尺。

　　罗伯特做好了坠机甚至更坏的心理准备。坚硬而冰冻的地表正扑面而来，那广阔而刺眼的雪地，在敌人追踪式火力的照耀下，发出红光。在第一次被击中几分钟后，机腹位置的金属便严重断裂。飞机向上攀升了片刻，便再次下坠，伴随着刺耳的声音，它猛地往黑暗的森林直冲而去。

　　这架战斗机厄运难逃，左翼卡在一根粗树干上，而机翼已经出现断裂了。在一阵剧烈的震动之后，它停了下来，半身埋到雪地里，机身前部则插入浓密的枝叶中。此时，罗伯特已经失去了知觉。①

　　刘易斯于2006年世界读书日获评美国"最受喜爱的二十位作家"之一，其作品被译为了三十多种语言。《飞行战犬》是英国作家、前战地记者刘易斯的代表作。该小说来源于第二次世界大战的真实事件，是一个有关残酷战争、历史、勇气以及人与动物真挚情感的令人难以置信的故事。故事记述的是一个其他国家的飞行员带着他发现并救下来的一只犬来到英国抗击德国纳粹，飞行战斗员和这只犬在战争中表现出了无畏勇气和忠诚，令人钦佩和感动，也令人难以置信。

　　上面这段描述的就是故事一开头就很惨烈，记述的是主人公的飞机被纳粹德国击落的整个事件过程。故事情节的翻译不同于景物描写和人物描写的翻译，故事情节更注重动作及事件的过程。景物描写和人物描写更注重人的主观感受，故事情节则更注重客观事实的描写。这一段就描写了第二次世界大战中一架被德国枪炮击中的飞机从被击中到发生失灵直至坠落的全过程。第一段第一句"Their controlled descent through the mist was over in a matter of seconds."在英语原文中的表达方式是静态的，中文译文译为"仅仅几秒钟后，他们便无法在雾中控制飞机的下降幅度。"将英文的静态转换为了中文的动态，同时将时间状语"仅仅几秒钟后"由句末提至句首，将英文的抽象的物的主语"descent"转换为人称主语"他们"，从而使句子的表达符合了中文的表达习惯和叙事视角，流畅清晰。第一段第二句"In spite of Pierre's desperate manoeuvres the German gunners quickly found their mark."译为了"尽管皮尔极力地变换飞行轨迹，他们还是被德国枪炮手发现了。"将英文原文中的抽象词"manoeuvres"具体化为了"飞行轨迹"，从而省去了后面的"mark"不译，避免了直译可能会产生的生涩之感。第二段第二句"The hard, frozen ground was rushing up to meet them, a wide expanse of glistening snow lit here and there a fiery red by the tracer fire."译为"坚硬而冰冻的地表正扑面而来，那广阔而刺眼的雪地，在敌人追踪式火力的照耀下，发出红光。"其中"The hard, frozen ground was rushing up to meet them"是较难翻译的，译者译为"扑面而来"，娓娓动听，较好地描

---

①　达米恩·路易斯.飞行战犬[M].李平,徐菊清,译.桂林:广西师范大学出版社,2017.

述出了飞机坠地刹那间的令人毛骨悚然的情景。整体来说,译文较好地传译出了飞机坠机事件的整个过程,明白晓畅。

外译中的重要作用是让我们了解世界,引进异质,从而发展我们自己的文明与文化,其作用甚大。季羡林曾指出:"不管经过了多少波折,走过多少坎坷的道路,既有阳关大道,也有独木小桥,中华文化反正没有消逝。原因何在呢? 我的答复是:倘若拿河流来作比,中华文化这一条长河,有水满的时候,也有水少的时候;但却从未枯竭。原因就是有新水注入。注入的次数大大小小是颇多的。最大的有两次,一次是从印度来的水,一次是从西方来的水。而这两次的大注入依靠的都是翻译。中华文化之所以能长葆青春,万应灵药就是翻译。翻译之为用大矣哉!"所以,国外航空航天文学的中文翻译对催生、发展我国的航空航天文学,特别是近年来我国享誉世界的科幻文学的发展,起到了重要作用。

练 习 题

一、基础练习:请翻译以下句子

1. I was descending a steep, cobbled, excavated road between banked-up footways, perhaps six feet high, upon which, in a monotonous series, opened the living room doors of rows of dark, low cottages.

2. The perspective of squat blue slate roofs and clustering chimneys drifted downward towards the irregular open space before the colliery—a space covered with coaly, wheel-scarred mud, with a patch of weedy dump to the left and the colliery gates to the right.

二、拓展练习:请翻译以下段落

Beyond, the High Street with shops resumed again in good earnest and went on, and the lines of the steam-tramway that started out from before my feet, and were here shining and acutely visible with reflected skylight and here lost in a shadow, took up for one acute moment the greasy yellow irradiation of a newly lit gas lamp as they vanished round the bend.

第三节 航空航天文学翻译:汉译英的原则

我们认为,航空航天文学汉译英的翻译原则有两个:一是在语言上求达,不求雅。这是由英文的文体特点决定的,英文讲究平实,不求典雅。这不同于航空航天文学的英译汉翻译原则;二是在内容上求神似,不求形似。这与航空航天文学的英译汉翻译原则相同。这同样意味着译文可根据目的语需求对原文的内容和语言形式进行适度的修改或增、省等翻译操作。下面,我们同样针对航空航天文学文本中的景物描写、人物描写及事件描写的翻译,通过具体实例来对语言和内容这两方面的翻译原则进行探讨。

一、景物描写的翻译

例7 中文原文:

在叶文洁的记忆中,这段日子不像是属于自己的,仿佛是从别的人生中飘落的片段,像一片羽毛般飞入自己的生活。这段记忆被浓缩成一幅幅欧洲古典油画,很奇怪,不是中国画,就是油画,中国画上空白太多,但齐家屯的生活是没有空白的,像古典的油画那样,充满着浓郁得化不开的色彩。一切都是浓烈和温热的:铺着厚厚乌拉草的火坑、铜烟锅里的关东烟和莫合烟、厚实的高粱饭、六十五度的高粱酒……但这一切,又都在宁静与平和中流逝着,像屯子边上的小溪一样。

最令叶文洁难忘的是那些夜晚。齐猎头儿的儿子到城里卖蘑菇去了,他是屯里第一个外出挣钱的人,她就和大凤住在一起。那时齐家屯还没通电,每天晚上,她们俩守在一盏油灯旁,叶文洁看书,大凤做针线活。叶文洁总是不自觉地将书和眼睛凑近油灯,常常刘海被烤得吱啦一下,这时她俩就抬头相视而笑。大凤从来没出过这事儿,她的眼神极好,借着炭火的光也能干细活儿。两个不到半周岁的孩子睡在她身边的炕上,他们的睡相令人陶醉,屋里能听到的,只有他们均匀的呼吸声。叶文洁最初睡不惯火炕,总是上火,后来习惯了,睡梦中,她常常感觉自己变成了婴儿,躺在一个人温暖的怀抱里,这感觉是那么真切,她几次醒后都泪流满面但那个人不是父亲和母亲,也不是死去的丈夫,她不知道是谁。[①]

英文译文:

In Ye's memory, these months seemed to belong to someone else, like a segment of another life that had drifted into hers like a feather. This period condensed in her memory into a series of classical paintings—not Chinese brush paintings but European oil paintings. Chinese brush paintings are full of blank spaces, but life in Qijiatun had no blank spaces. Like classical oil paintings, it was filled with thick, rich, solid colors. Everything was warm and intense: the heated kang stove-beds lined with thick layers of ura sedge, the Guandong and Mohe tobacco stuffed in copper pipes, the thick and heavy sorghum meal, the sixty-five-proof baijiu distilled from sorghum—all of these blended into a quiet and peaceful life, like the creek at the edge of the village.

---

① 刘慈欣. 三体[M]. 重庆:重庆出版社,2008.

Most memorable to Ye were the evenings. Hunter Qi's son was away in the city selling mushrooms—the first to leave the village to earn money elsewhere, so she shared a room in his house with Feng. Back then, there was no electricity in the village, and every evening, the two huddled around a kerosene lamp. Ye would read while Feng did her needle work. Ye would lean closer and closer to the lamp without noticing, and her bangs would often get singed, at which point the two of them would glance up and smile at each other. Feng, of course, never had this happen to her. She had very sharp eyes, and could do detailed work even in the dim light from heating charcoal. The two babies, not even half a year old, would be sleeping together on the kang next to them. Ye loved to watch them sleep, their even breathing the only sound in the room.

At first, Ye did not like sleeping on the heated kang, and often got sick, but she gradually got used to it. As she slept, she would imagine herself becoming a baby sleeping in someone's warm lap. The person who held her wasn't her father or mother, or her dead husband. She didn't know who it was. The feeling was so real that she would wake up with tears on her face.[①]

这一段是刘慈欣的《三体》中叶文洁对居住在大兴安岭齐家屯的一段时间的生活描写。在语言方面，我们看到译者刘宇昆的英文译文风格非常平实，没有运用典雅之词，朴实无华。中文原文中第一段的描述是颇有诗意的，刘宇昆的英语译文基本是对照翻译的，主要是做到了"达"的要求。"像古典的油画那样，充满着浓郁得化不开的色彩"译为了"Like classical oil paintings, it was filled with thick, rich, solid colors."，化解了中文原文中的夸张的表达方式，转换为了平实的英文风格，并没有像原文那样雅致。对于第二段的翻译，刘宇昆将其切分为了两段，主要是因为话题发生了转换，这样翻译的主要目的也是为了英语叙事上的"达"的要求。

例8 中文原文：

我们很快到达了海边，看到城市摩天大楼的尖顶伸出海面，退潮时白花花的海水从大楼无数的窗子中流出，形成一道道瀑布……刹车时代刚刚结束，其对地球的影响已触目惊心：地球发动机加速造成的潮汐吞没了北半球三分之二的大城市，发动机带来的全球高温融化了极地冰川，更给这大洪水推波助澜，波及南半球。爷爷在三十年前亲眼看见了百米高的巨浪吞没上海的情景，他现在讲这事的时候眼还直勾勾的。事实上，我们的星球还没启程就已面目全非了，谁知道在以后漫长的外太空流浪中，还有多少苦难在等着我们呢？我们乘上一种叫船的古老的交通工具在海面上航行。

---

① Liu Cixin. The Three-body Problem[M]. Trans. by Ken Liu. New York: Tor Books, 2016.

地球发动机的光柱在后面越来越远,一天以后就完全看不见了。这时,大海处在两片霞光之间,一片是西面地球发动机的光柱产生的青蓝色霞光,一片是东方海平面下的太阳产生的粉红色霞光,它们在海面上的反射使大海也分成了闪耀着两色光芒的两部分,我们的船就行驶在这两部分的分界处,这景色真是奇妙。但随着青蓝色霞光的渐渐减弱和粉红色霞光的渐渐增强,一种不安的气氛在船上弥漫开来。甲板上见不到孩子们了,他们都躲在船舱里不出来,舷窗的帘子也被紧紧拉上。[①]

英文译文:

After a while, we made our way to the ocean. Standing on the seashore, we could see the pinnacles of submerged skyscrapers reaching up out of the waves with the ebb of the tide. We beheld the gleaming white wash of water rush out of their windows, forming cascades of waterfalls.

Back then the Reining Age had only just come to an end, leaving the Earth with the horrifying aftermath of its passing. The tides, quickened by the Earth Engines, had swallowed two out of every three cities in the Northern Hemisphere; then the global increase in temperatures melted the polar icecap, turning the ensuing floods into a deluge that spread to the Southern Hemisphere. Thirty years earlier, my grandfather had witnessed giant 300-foot waves that had engulfed Shanghai. Even now, he could never tell us about it without his gaze slipping into a thousand-mile stare.

Our planet had already changed beyond recognition before it even set out on its journey. Who knew what hardships awaited us on our long travels through outer space?

At the seashore we boarded an archaic vessel called "ship". As we departed the coast, the Earth Engines grew ever more distant. Within a day's travel, they had disappeared altogether behind us. Before us the ocean was bifurcated by light; in the west, the azure glow of the Earth Engines' jets; in the east, the shimmering pink water, illuminated by the Sun's rays. We sailed straight down the glittering seam where the two glows met on the ocean's surface. It was a truly marvelous sight to witness. As our voyage continued, the azure glow slowly waned, while the pink light gradually waxed. With its waxing, unease began to spread across the ship. We children could no longer be seen on deck. Seeking shelter in the belly of the ship, we even drew the porthole blinds tight.[②]

---

① 刘慈欣. 流浪地球[M]. 武汉:长江文艺出版社,2008.
② Liu Cixin. The Wandering Earth[M]. Trans. by Holger N. New York: Tor Trade, 2022.

这是《流浪地球》中描述的地球进入刹车时代后人类进入太空的情景。我们首先看到的译文与原文相比的变化之处是：原文为一段，但译文分为了四段。因此，在中文文学作品英译中，段落的重构是一个重要的问题。但整体来说，原文译文基本上也是中文原文的对照翻译，主要是做到了"达"。与刘宇昆英译的《三体》不同的是，译者Holger Nahm 在用词上更有英语的特点。例如，将第一句"我们很快到达了海边，看到城市摩天大楼的尖顶伸出海面，退潮时白花花的海水从大楼无数的窗子中流出，形成一道道瀑布。"译成"After a while, we made our way to the ocean. Standing on the seashore we could see the pinnacles of submerged skyscrapers reaching up out of the waves with the ebb of the tide. We beheld the gleaming whitewash of water rush out of their windows, forming cascades of waterfalls."，用上了一些比较地道的英文语汇，如 pinnacle, whitewash, beheld, cascade 等。

例 9　中文原文：

爆炸的情景我是亲眼看到的。我们听到一声巨响时都紧急地回头或抬头看河堤，这时尚未爆炸，艇头撞起来的泥沙正在下落，它的两扇巨翅和翘起来的尾巴疯狂地抖动着。紧接着就爆炸了。

我们首先看到一团翠绿的强光在河堤上凸起，绿得十分厉害，连太阳射出的红光都被逼得弯弯曲曲。随着绿光的凸起，半条河堤都突然扭动起来。成吨的黑土翻上了天。这时候我们才听到一声沉闷的轰响，声音并不是很大，好像从遥远的旷野里传来的一声狮吼。我后来才知道"大音稀声"的道理。这一声爆炸方圆四十里都能听到，不知有多少人家的窗户纸都给震破了。几乎与听到轰响同时，我感到脚下的道路在跳动。路边的白杨树枝哗啦啦地响着，方家七老妈像神婆子跳大神一样跳跃着。

我们扔掉的要饭篮也在地上翻滚着。我看到我们的叫花子队伍像谷个子一样翻倒了，我在感觉着上边那些景象的同时，胸前仿佛被一只无形的巨掌猛推了一下子。我恍恍惚惚地看到无垠的天空上流动着鸢尾花的颜色，漂亮又新鲜，美好又温柔。
（莫言，《飞艇》）

英文译文：

I saw how the flying ship exploded with my own eyes. When we heard the terrific noise, we all turned or looked up quickly toward the embankment. The flying ship hadn't exploded yet and the earth spewed up by the nose of the thing was still raining down. Its giant wings and its raised tail were shuddering like crazy. And right after that, it exploded.

First we saw a ball of strong green light rear up out of the embankment. It was fiercely green, it even bent the red rays of the sun out of shape. Following the rise of the green light, half the embankment was abruptly set in motion and tons of black earth flung skyward. And then we heard a dull boom. The noise wasn't that loud, it was like a lion roaring in a far-off wilderness. Only later did I learn Zhuangzi's principle that "a great noise makes a small sound". The explosion could be heard over a twenty kilometre radius. I wonder how many paper windows it tore. At practically the same time as I heard the explosion, I felt the road quake under my feet. The branches of the poplar trees along the road shivered and Granny Fang hopped around like a woman possessed.

The discarded begging baskets rolled around on the ground too. I saw our little platoon of beggars collapse like bundles of millet, and at the same time as I sensed what I've described above, some amorphous giant fist suddenly gave me a shove in the chest. As if in a trance, I saw the endless reaches of space flowing iris-hued, lovely, vibrant and warm.（葛浩文译，*The Flying Ship*）

这是莫言在其短篇小说《飞艇》中对一场飞艇爆炸的场面的描述。我们看到，葛浩文的英译达到了"达"的要求。对第二段中"大音希声"的翻译中，添加了相关文化背景信息，译为了"Zhuangzi's principle that 'a great noise makes a small sound'"，并将"方家七老妈像神婆子跳大神一样跳跃着"这种我国文化特色的表达处理成了英语中易懂的表达方式：Granny Fang hopped around like a woman possessed。

二、人物描写的翻译

例 10 中文原文：

清晨 4:50，老刀穿过熙熙攘攘的步行街，去找彭蠡。

从垃圾站下班之后，老刀回家洗了个澡，换了衣服。白色衬衫和褐色裤子，这是他唯一一套体面衣服，衬衫袖口磨了边，他把袖子卷到胳膊肘。老刀四十八岁，没结婚，已经过了注意外表的年龄，又没人照顾起居，这一套衣服留着穿了很多年，每次穿一天，回家就脱了叠上。他在垃圾站上班，没必要穿得体面，偶尔参加谁家小孩的婚礼，才拿出来穿在身上。这一次他不想脏兮兮地见陌生人。他在垃圾站连续工作了五小时，很担心身上会有味道。

步行街上挤满了刚刚下班的人。拥挤的男人女人围着小摊子挑土特产，大声讨价还价。食客围着塑料桌子，埋头在酸辣粉的热气腾腾中，饿虎扑食一般，白色蒸汽遮住了脸。油炸的香味弥漫。货摊上的酸枣和核桃堆成山，腊肉在头顶摇摆。这个点是全天最热闹的时间，基本都收工了，忙碌了几个小时的人们都赶过来吃一顿饱饭，人声鼎沸。①

英文译文：

At ten of five in the morning, Lao Dao crossed the busy pedestrian lane

---

① 郝景芳.北京折叠[M].南京：江苏凤凰文艺出版社，2016.

on his way to find Peng Li.

After the end of his shift at the waste processing station, Lao Dao had gone home, first to shower and then to change. He was wearing a white shirt and a pair of brown pants—the only decent clothes he owned. The shirt's cuffs were frayed, so he rolled them up to his elbows. Lao Dao was forty-eight, single, and long past the age when he still took care of his appearance. As he had no one to pester him about the domestic details, he had simply kept this outfit for years. Every time he wore it, he'd come home afterward, take off the shirt and pants, and fold them up neatly to put away. Working at the waste processing station meant there were few occasions that called for the outfit, save a wedding now and then for a friend's son or daughter.

Today, however, he was apprehensive about meeting strangers without looking at least somewhat respectable. After five hours at the waste processing station, he also had misgivings about how he smelled.

People who had just gotten off work filled the road. Men and women crowded every street vendor, picking through local produce and bargaining loudly. Customers packed the plastic tables at the food hawker stalls, which were immersed in the aroma of frying oil. They ate heartily with their faces buried in bowls of hot and sour rice noodles, their heads hidden by clouds of white steam. Other stands featured mountains of jujubes and walnuts, and hunks of cured meat swung overhead. This was the busiest hour of the day—work was over, and everyone was hungry and loud.①

这一段是郝景芳的《北京折叠》中一开头对老刀这个人物的出场及其周围人物的描写。老刀，单身，在垃圾站上班，周围的人物也是乱哄哄、熙熙攘攘。译者刘宇昆将第一段进行了切分，将"Today, however, he was apprehensive about meeting strangers without looking at least somewhat respectable. After five hours at the waste processing station, he also had misgivings about how he smelled."这部分单独成段，因话题发生了转换。译者也基本上是对原文进行照译，译文基本通达，但并不出彩。

例 11　中文原文：

地球三体叛军并非铁板一块，它的内部有着复杂的派别和分支，主要分为两部分：

降临派：这是三体叛军最本原最纯粹的一脉，主要由伊文斯物种的信奉者组成。

---

① Hao J F. Folding Beijing[M]. Trans. by Ken Liu. London：Head of Zeus, 2017.

他们对人类本性都已彻底绝望,这种绝望最初来源于现代文明导致的地球物种大灭绝,伊文斯就是其典型代表。后来,降临派对人类的憎恨开始有了不同的出发点,并非只局限于环保和战争等,有些上升到了相当抽象的哲学高度。与后来人们的想象不同,这些人大都是现实主义者,对于他们为之服务的外星文明也并未抱太多的期望,他们的背叛只源于对人类的绝望和仇恨,麦克伊文斯的一句话已成为降临派的座右铭:我们不知道外星文明是什么样子,但知道人类。

拯救派:这是在三体叛军出现相当长的时间后才产生的一个派别,它本质上是一个宗教团体,由三体教的教徒组成。①

英文译文:

The members of the ETO were not of a single mind. Within the organization were complicated factions and divisions of opinion. Mainly, they fell into two factions.

The Adventist group was the purest, most fundamentalist strand of the ETO, comprised mainly of believers in Evans's Pan-Species Communism. They had completely given up hope in human nature. This despair began with the mass extinctions of the Earth's species caused by modern civilization. Later, other Adventists based their hatred of the human race on other foundations, not limited to issues such as the environment or warfare. Some raised their hatred to very abstract, philosophical levels. Unlike how they would be imagined later, most of them were realists, and did not place too much hope in the alien civilization they served either. Their betrayal was based only on their despair and hatred of the human race. Mike Evans gave the Adventists their motto: We don't know what extraterrestrial civilization is like, but we know humanity.

The Redemptionists didn't appear until long after the ETO's founding. This group's nature was a religious organization, and the members were believers in the Trisolaran faith.②

这是《三体》第二十九章关于地球三体运动中对地球三体叛军的描写。因为这是一部科幻小说,许多人物或组织都是现实社会中所找不到的,自然这就要求译者进行创造性翻译,寻求神似。在这方面,译者刘宇昆就进行了很好的创译。例如,将"地球三体运动"和"地球三体叛军"译为了"Earth-Trisolaris Movement",简称"ETO",模仿了已有的"EFO",颇有科幻之感;"降临派"和"拯救派"译为了"Adventist"和"Redemptionist",既保留了中文原意,又在英文中可理解,可谓"神似"之译。

---

① 刘慈欣. 三体[M]. 重庆:重庆出版社,2008.
② Liu Cixin. The Three-body Problem[M]. Trans. by Ken Liu. New York:Tor Books, 2016.

三、故事情节的翻译

例 12　中文原文：

晨光熹微中，一座城市折叠自身，向地面收拢。高楼像最卑微的仆人，弯下腰，让自己低声下气切断身体，头碰着脚，紧紧贴在一起，然后再次断裂弯腰，将头顶手臂扭曲弯折，插入空隙。高楼弯折之后重新组合，蜷缩成致密的巨大魔方，密密匝匝地聚合到一起，陷入沉睡。然后地面翻转，小块小块土地围绕其轴，一百八十度翻转到另一面，将另一面的建筑楼宇露出地表。楼宇由折叠中站立起身，在灰蓝色的天空中像苏醒的兽类。城市孤岛在橘黄色晨光中落位，展开，站定，腾起弥漫的灰色苍云。①

英文译文：

In the early dawn, the city folded and collapsed. The skyscrapers bowed submissively like the humblest servants until their heads touched their feet; then they broke again, folded again, and twisted their necks and arms, stuffing them into the gaps. The compacted blocks that used to be the skyscrapers shuffled and assembled into dense, gigantic Rubik's Cubes that fell into a deep slumber.

The ground then began to turn. Square by square, pieces of the earth flipped 180 degrees around an axis, revealing the buildings on the other side. The buildings unfolded and stood up, awakening like a herd of beasts under the gray-blue sky. The island that was the city settled in the orange sunlight, spread open, and stood still asmisty gray clouds roiled around it.②

这一段是郝景芳的《北京折叠》中对折叠城市如何进行折叠的故事情节的描述。虽有一些拟人修辞手法的运用，但基本上都是对城市折叠这个事件的客观描述。刘宇昆的英文译文也较为清晰地描述出了这个城市折叠的过程，基本以"达"为目的。

例 13　中文原文：

我们一出村头，就看到从南边飞出来了。太阳刚出，状如盛粮食的大囤，血红的颜色，洇染了地平线和低空中的云彩。遍野的枯草茎上，挂着刺刺茸茸的白霜。路上龟裂着多叉的纹路。在很远的地方发出过一阵如雷的轰鸣，在原野上滚动。临近我们村庄时，却突然没有了声息。那时候我们都站在村头那条通向南山的灰白道路上，我们挎着讨饭篮，挂着打狗棍（吓狗棍，绝对不能打人家的狗），看到银灰色的飞艇从几百米的空中突然掉下来，掉到离地五六十米高时，它斜着翅膀子，哆哆嗦嗦往前飞，不是飞，是滑翔！我听到的肚子里噼里啪啦地响着，两股浓密的黑烟从翅膀后冒出来，拖得很长，好像两条大尾巴。擦着路边的白杨树梢滑过去，直扑着我们的村庄去

---

① 郝景芳.北京折叠[M].南京：江苏凤凰文艺出版社，2016.
② Hao J F. Folding Beijing[M]. Trans. by Ken Liu. London: Head of Zeus, 2017.

了。虽然机器不响,但仍然有尖利的呼啸,白杨树上的枯枝嚓啦啦响着,树上的喜鹊和乌鸦一齐惊飞起来。强劲的风翻动着我们破烂的衣衫。方家七老妈前走走,后倒倒,好像随时要倒地。像一个巨大的阴影从地上飞掠而过。我们都胆战心惊,每个人都表现出了自己的最丑陋的面容。连姐姐的搽过雪花膏的脸蛋也惨不忍睹。姐姐惊愕得大张着嘴巴,额头上布满横一道竖一道的皱纹。我是期望着降落到我们村庄里去的,但是它偏不,它本来是直冲着我们的村庄扎下去了,它的肚皮拉断了方六老爷家一棵白杨树的顶梢,一颗像轧场的碌碡那么粗的、乌溜溜闪着蓝光的、屁股上生着小翅膀的可爱的玩意儿掉在我们生产队的打谷场上。后来才知道那是颗大炸弹。拉断了一棵树,又猛地昂起头,嘎嘎吱吱地拐了一个弯,摇摇晃晃,哆哆嗦嗦,更像个醉鬼,掉头向东来了。它的翅膀上涂满了阳光,好像流淌着鲜血。这时它飞得更低了,速度也更快,体形也更大,连里面的三个人都能看清楚,他们的脸都是血红的。它的巨翅像利剑一样从我们头上削过去,我们都捂住脑袋,在这样的情况下,没有一个人感到自己的头颅是安全的。(莫言,《飞艇》)

英文译文:

As soon as we left the village, we saw the flying ship coming up from the south. The sun was just up and it looked like a huge grain basket, blood red, staining the horizon and the low clouds. Everywhere, sharp frost covered the dried stubble of grass like down. The road was chapped and ragged from the cold. From the far-off flying ship a thunderous roar rolled across the plain. But when it drew near our village there was a sudden silence. We were standing on the grey road leading out of our village toward Nanshan, begging baskets in hand, dog-beating poles over our shoulders (dog-scaring, that is; we beggars certainly couldn't beat someone else's dog). We saw the silver-grey flying ship suddenly fall out of the sky from several hundred metres up. When it had fallen to about fifty or sixty metres, the wings tilted and it shivered onwards for a bit. No, it wasn't flying, it was gliding! I heard some rumbling and clattering coming from the belly of the thing and two columns of thick black smoke spewed from behind the wings, trailing out like two big tails. The flying ship skimmed over the tops of the white poplars along the road, heading straight for our village. Though the engine was silent, there was still a shrill whistling that made the dry branches of the poplars crackle and the magpies and crows in them take flight in startled unison. A strong wind ruffled our tattered clothes. Granny Fang stumbled backwards and forwards a bit, as if she might fall over any minute. The flying ship swept overhead like a giant shadow, the flying ship's giant shadow swept across the ground. We were all terror-stricken, every face wearing its ugliest expression. Even my sister's, with its coating of face-cream, was too hideous to look at, with her mouth wide open in shock and the skin on her forehead standing out in ridges. I was hoping the flying ship would come down in our village. But it didn't. At first it was hurtling straight toward it, but then its belly tore the top of a poplar tree at Sixth Grandpa Fang's and a neat little gizmo about the size and shape of a millstone, shiny black with blue glints and little wings sprouting from its rear end fell onto the production team's threshing floor. We later learned that it was a bomb. After ripping the top off the tree, the flying ship suddenly raised its nose and rattled into a turn, teetering for all the world like someone who'd had a few too many, and turned to head east toward us. Its wings, smeared with sunlight, looked as if they were gushing fresh blood. It was flying even lower now, and had picked up speed. It was bigger than ever and we had a clear view of the three men inside. Their faces were pure red. The huge wings of the flying ship sliced through the air above us like a sharp sword and we all covered our heads with our hands. In the circumstances, we all feared for our safety. (葛浩文译, *The Flying Ship*)

这是莫言的短篇小说《飞艇》中一段描写飞艇袭击村庄的情节。莫言的语言文学性较强,夸张性、形象性较强。自然,翻译起来,难度就更大。葛浩文的英语译文同样文笔精彩,传神达意。例如,莫言使用的这个"掉到离地五六十米高时,它斜着翅膀子,哆哆嗦嗦往前飞,不是飞,是滑翔! 我听到的肚子里噼里啪啦地响着,两股浓密的黑烟从翅膀后冒出来,拖得很长,好像两条大尾巴。"这个拟人修辞手法是较难翻译的,但葛浩文却轻松化解,将其译为"When it had fallen to about fifty or sixty metres, the wings tilted and it shivered onwards for a bit. No, it wasn't flying, it was gliding! I heard some rumbling and clattering coming from the belly of the thing and two columns of thick black smoke spewed from behind the wings, trailing out like two big tails.",可谓有异曲同工之妙。再如,描述大家惊恐之状的"我们都胆战心惊,每个人都表现出了自己的最丑陋的面容。连姐姐的搽过雪花膏的脸蛋也惨不忍睹。姐姐惊愕得大张着嘴巴,额头上布满横一道竖一道的皱纹。"这句话也是较难翻译的,如"横一道竖一道的皱纹"。葛浩文的译文并没有仅仅拘泥于字面,而是将其灵活翻译为"We were all terror-stricken, every face wearing its ugliest expression. Even my sister's, with its coating of face-cream, was too hideous to look at, with her mouth wide open in shock and the skin on her forehead standing out in ridges.",传达出了原作的神韵。

在我国,到了清末民初,作为新民和启蒙的有效手段,小说的社会功能被提高到前所未有的高度,甚至被神化:欲新一国之民,不可不先新一国之小说。故欲新道德,必新小说;欲新宗教,必新小说;欲新政治,必新小说;欲新风俗,必新小说;欲新学艺,必新小说;乃至欲新人心,欲新人格,必新小说。何以故？小说有不可思议之力支配人道故。①

因此,小说的翻译在现当代文学翻译中占有重要的位置,特别是近年来我国科幻小说的英译更是成了我国文学外译的意外的风景线,大大提升了我国现当代文学在世界文学中的地位。当然,我们也看到,我国小说的翻译水平也还参差不齐,有的仅仅做到了达意,有的则不仅达意而且还传神。因此,我们还需进一步提升我国当代文学翻译的水平,从文学翻译走向翻译文学。

练 习 题

一、基础练习:请翻译以下句子

1. 老刀从来没有见过这样的景象。太阳缓缓升起,天边是深远而纯净的蓝,蓝色下沿是橙黄色,有斜向上的条状薄云。

2. 太阳被一处屋檐遮住,屋檐显得异常黑,屋檐背后明亮夺目。

3. 太阳升起时,天的蓝色变浅了,但是更宁静透彻。

4. 他想抓住那道褪去的金色。蓝天中能看见树枝的剪影。他的心狂跳不已。他从来不知道太阳升起竟然如此动人。

二、拓展练习:请翻译以下段落

他跑了一段路,停下来,冷静了。他站在街道中央。路的两旁是高大的树木和大片草坪。他环视四周,目力所及,远远近近都没有一座高楼。他迷惑了,不确定自己是不是真的到了第一空间。他能看见两排粗壮的银杏。

---

① 梁启超. 论小说与群治之关系 1902/1997[C]∥二十世纪中国小说理论资料.北京：北京大学出版社,1997：1887-1916.

第三章 航空航天文学翻译的策略与方法：英译汉

第一节 航空航天文学翻译英译汉的策略

"翻译策略"是在翻译活动中，译者为了实现特定的翻译目的所采用的一系列原则和方案的集合[①]。它强调的是宏观上的整体框架，对译文的整体效果会产生巨大影响。翻译策略大致可以分为"异化"和"归化"两类。

一、异 化

"异化"(foreignization)是指译者在翻译过程中，尽可能地保留原语文本的表达方式。这种策略具有原文作者倾向，在形式上译文靠近原文，一般能够更加准确地传达出原文的意思。如下例：

例1 英文原文：

It also feels as though a huge truck going at top speed just smashed into the side of us. Perfectly normal, apparently, and we'd been warned to expect it. So I just keep "hawking it," flipping through my tables and checklists and staring at the buttons and lights over my head, scanning the computers for signs of trouble, trying not to blink.[②]

中文译文：

那感觉还好像一辆疾驰的大卡车一下子撞上我们的侧面。很显然，这些感觉都完全正常，我们之前得到过提醒，知道会有这样的感觉。所以我只能是"像鹰一样"机警地观察着，翻阅着我的表格和检查清单，盯着头上的按钮和指示灯，审视着电脑看有没有出问题的迹象，连眼睛都不敢眨一下。[③]

本例中，划线部分的 hawk 源自短语 watch someone like a hawk，根据柯林斯词典的释义，可以知道这一表达的意思是"(像老鹰一样)仔细地关注"，译者保留了原文中的特色表达，一方面非常形象贴切地描述出人物的状态，同时使译文读起来多了

[①] 熊兵. 翻译研究中的概念混淆：以"翻译策略"、"翻译方法"和"翻译技巧"为例[J]. 中国翻译，2014(3)：82-88.

[②] Chris H. An Astronaut's Guide to Life on Earth[M]. Toronto：Random House Canada，2013.

[③] 哈德菲尔德. 宇航员地球生活指南[M]. 徐彬，译. 长沙：湖南科学技术出版社，2017.

几分异域色彩。

语言形式为意义传达服务,特定的语言形式可以体现独特的语言及文化特色。采用异化策略能够较好地保留这些因素,同时让读者了解到新颖的表达方式,给目的语言带来活力。但是,过度保留原语的表达方式会降低译文语言的自然性和可读性,影响目标读者的接受程度。因此,运用异化策略具有一定的限度,一味地异化会使译文晦涩难懂,甚至影响作者真正意义的传达。译者在翻译时需要考虑到目的语的接受程度,以及目标读者是否能够理解并接受,从而灵活选择。

二、归 化

"归化"(domestication)是指译者在翻译中尽可能地采用接近于目的语习惯的表达方式,从而使翻译结果更加符合目的语读者的阅读习惯。这种策略具有译文接受者倾向,译文易于读者理解和接受。请看下例:

例 2 英文原文:

There on the one hand in a crowded darkness, about the ugly factories and work-places, the workers herded together, ill clothed, ill nourished, ill taught, badly and expensively served at every occasion in life, uncertain even of their insufficient livelihood from day to day, the chapels and churches and public-houses swelling up amidst their wretched homes like saprophytes amidst a general corruption, and on the other, in space, freedom, and dignity, scarce heeding the few cottages, as overcrowded as they were picturesque, in which the laborers festered, lived the landlords and masters who owned pot-banks and forge and farm and mine.①

中文译文:

在那里,一方面,在熙熙攘攘的黑暗中,在那些简陋的工厂和车间附近,工人们成群结队,破衣烂衫,营养不良,目不识丁,生活中处处收费昂贵还要遭受白眼,日复一日,吃了上顿没下顿,那些教堂和小旅馆在他们肮脏破烂的住房中间拔地而起,真像是一大片腐烂变质物体上冒出的腐生植物。另一方面,自由和尊贵难得光顾的为数不多的村舍则过分拥挤,一派如梦似画的景象。工人们在这里苦苦挣扎,怨声载道,有钱人却过得逍遥自在,他们拥有瓷器厂、铁匠铺、农场和矿井。②

本例中,原文采用了较为复杂的长句,通过对细节的生动描述展现了两种截然不同的生活环境。译文对句子进行了切分和调整,使之成为相对短小的句子,更加符合中文读者的阅读节奏。此外,译者采用了不少四字结构,如"破衣烂衫""目不识丁""如梦似画""怨声载道"等,更贴近中国读者的语言习惯和审美体验。同时,对于一些

---

① Herbert G W. In the Days of the Comet[M]. Auckland: The Floating Press, 2009.
② 赫伯特·乔治·威尔斯. 彗星来临[M]. 穆雷,译. 大连:大连理工大学出版社,2018.

相对抽象的形容词,译者也进行了解释,如将"uncertain"转换为"吃了上顿没下顿",容易被读者理解。

不过,归化使译文在形式上远离原文,译文的忠实度和准确度可能会有所下降,有时无法充分且完整地传达作者的风格和特色。因此,归化也应当有一定的限度,不恰当的过度归化,忽略原语的文化特征,单纯追求通顺流畅,会产生极大的误导影响。

例3　英文原文:

Between them would be placed a bottle of Janx Spirit (as immortalized in that ancient Orion mining song, "Oh, don't give me none more of that Old Janx Spirit/ No, don't you give me none more of that Old Janx Spirit/For my head will fly, my tongue will lie, my eyes will fry and I may die/Won't you pour me one more of that sinful Old Janx Spirit").①

中文译文:

两人之间搁着一瓶"销魂浆",这好酒声名远播,有古老的猎户座采矿歌赞曰:

销魂琼浆莫多饮

多饮头昏吐狂言

销魂琼浆莫多饮

多饮目裂人归西

将进酒杯莫停

销魂琼浆催人罪②

本例中,英语原文介绍了一种喝酒游戏,并且提到赞美这种烈酒的一首采矿歌。译文将酒的名字译为"销魂浆",而且用古代汉语诗歌的形式翻译歌词内容。原文中的人物仿佛像李白一样吟诗饮酒,使译文读起来失去了原本的风格,给读者一种格格不入、时空错乱的感觉。

异化与归化的概念并不是绝对的,体现的是译者在翻译选择上的倾向性。纯粹的异化或归化在翻译实践中并不常见,有的译本以异化为主,有的译本以归化为主。在实际翻译中,译者往往会在异化和归化之间进行权衡,并结合翻译文本的特点、出版社的要求等选择最合适的翻译策略。

例4　英文原文:

Following the meeting, we bought a drawing board, some T-squares, triangles, and other drafting equipment, and headed back to our quarters in Mayfair Court.③

---

① Douglas A. The Hitchhiker's Guide to the Galaxy[M]. New York: Ballantine Books, 2004.
② 道格拉斯·亚当斯. 银河系搭车客指南[M]. 姚向辉,译. 上海:上海译文出版社,2011.
③ Johnson C, Maggie S. Kelly: More Than My Share of It All[M]. Washington D. C.: Smithsonian Books, 1985.

中文译文：

洽谈之后，接着我们就买了一张绘图板、几把丁字尺、三角板以及其他绘图仪器设备，回到我们下榻的坐落在梅费尔街的寓所。①

本例中，在描述绘图仪器时，在两种选择中，译者没有将 T-squares 译为"T 形尺"，而采用与其形状相似的汉字，译为"丁字尺"，说明译者更倾向于采取归化策略，使译文更接近读者的知识背景。

航空航天文学中往往包含大量的科技专业术语，如机型名称、航空器部件、飞行术语等，译者需要准确译出这些专业表达。对于这些内容，译者通常采用异化的翻译策略，可以避免影响表达的准确性，产生歧义。同时，作为一种文学形式，这些作品也通常具有明显的文学特征，语言生动，表现力强。比如会涉及各种丰富的场景描写，还有不同性格或者身份人物之间的对话。例如：

例 5　英文原文：

Macie Roberts was careful when he said he needed to work with a computer. "You will not go to his office," she would say. "He only comes to our room to work. Don't meet him anywhere else." Barbara was curious; why couldn't they go to his office? She asked another girl, who laughed and then quietly said, "I'll show you later."②

中文译文：

当他提出要找一位计算员来协助工作的时候，梅茜·罗伯茨十分警惕。"你们不能去他的办公室，"她这样叮嘱手下的女孩，"真有事就让他上我们的办公室来。你们不能去其他任何地方见他。"芭芭拉十分好奇，为什么她们就不能去他的办公室呢？她问了另一个女孩，后者哈哈大笑，然后悄悄告诉她："一会儿我带你去看。"③

本段为年轻女子之间的对话，气氛轻松活跃，因此在翻译时，需要使译文对话具有相似的风格。比如 He only comes to our room to work 直译应当是"他只能来我们办公室工作"，但这样显得对话十分严肃正式，译文中"真有事就让他上我们的办公室来"很好地还原了原文中轻松愉悦的氛围和语气。同时，将 laugh 译为"哈哈大笑"，将 I will show you later 译为"我一会儿带你去看"都非常符合说话人的年龄、身份和性格。翻译人员需要结合具体文本，综合考虑如何处理各种要素，以达到最佳的翻译效果。

译者主要采取哪一种策略还会受到翻译活动参与者的影响，包括原文作者、翻译活动的发起人/委托人、译者本身、译文接受者等。比如，原文作者可能希望译者采取

---

① 凯利·约翰逊，玛吉·史密斯. 我怎样设计飞机：美国飞机设计师凯利·约翰逊自传[M]. 程不时，等译. 北京：航空工业出版社，1990.
② Holt N. Rise of the Rocket Girls[M]. New York：Little, Brown and Company, 2016.
③ 娜塔莉娅·霍尔特. 让火箭起飞的女孩[M]. 阳曦，译. 北京：九州出版社，2022.

更为异化的策略,确保自己的思想理念能被完整传达。而出版社可能希望译者采取更为归化的策略,来迎合大多数目的语读者的阅读习惯和偏好,使读者更易于理解和接受译文,使作品在目的语文化中得到更好的传播。译者本身的社会身份、翻译理念、价值观念等也会极大地影响他们使用的策略,影响译文的表达效果。对于译文接受者来说,不同的知识背景、阅读习惯、审美倾向会影响他们的阅读选择,对于原语文化了解不多的读者更倾向于选择归化程度更高的译本,而希望了解原语文化特色的读者更倾向于选择异化程度更高的译本。因此,译文的目标读者不同,译文也可能存在差异。

译者采用哪一种倾向有时从译作的前言或后记中便可见一斑。在 *Kelly: More Than My Share of It All*(《我怎样设计飞机》)一书的中译本前言中,译者明确提到"约翰逊自幼来自农村,他成为知名的飞机设计师之后,生活条件发生了很大的变化……虽然他声称自己生活的一些基本因素并没有改变,但是在行文之中也可以看出他对老年时富裕的生活条件流露出满足,甚至作了一些夸大的渲染",然而,由于这本自传记录了作者研制飞机过程中的亲身经历,"对于关心航空技术发展史的读者……是很有价值的第一手资料",因而"按原文译出未作删节",译文整体以异化策略为主,尽可能地保留原文中的表达方式和语言风格,保证内容的准确性。*The Science of Interstellar*(《星际穿越》)一书的译者序中提到,译者均为来自国家天文台的科研工作者,该书翻译"以保证科学内容的准确性为第一原则,语言风格力求与原文一致。"

练 习 题

一、基础练习:请翻译以下句子

1. The old dam we interrupted at her feeding on the Green Hill swept Balmy with a cold glance, kicked up her heels, and trotted toward the centre of the herd flinging over her shoulder at her spindly-legged foal a command to follow. But the foal never moved.

2. The wind in the wires is like the tearing of soft silk under the blended drone of engine and propeller.

3. Time and distance together slip smoothly past the tips of my wings without sound, without return, as I peer downward over the night-shadowed hollows of the Rift Valley and wonder if Woody, the lost pilot, could be there, a small human pinpoint of hope and of hopelessness listening to the low, unconcerned song of the Avian—flying elsewhere.

4. Everywhere on earth that day, in the ears of every onewho breathed, there had been the same humming in the air, the same rush of green vapors, the crepitation,

the streaming down of shooting stars.

二、拓展练习：请翻译以下段落

1. The Hindoo had stayed his morning's work in the fields to stare and marvel and fall, the blue-clothed Chinaman fell headforemost athwart his midday bowl of rice, the Japanese merchant came out from some chaffering in his office amazed and presently lay there before his door, the evening gazers by the Golden Gates were overtaken as they waited for the rising of the great star.

2. As I flew, my hunch became conviction. Nothing in the world, I thought, could have looked so much like reflecting water as the wings of Woody's plane. I remembered how bright those wings had been when last I saw them, freshly painted to shine like silver or stainless steel.

第二节 航空航天文学翻译英译汉的方法

"翻译方法"是译者在翻译活动中基于某种翻译策略，为达到特定翻译目的所采取的特定途径、步骤、手段[1]。由于两种语言之间存在差异，译者在翻译中会遇到各种困难，这时就需要分析应如何遣词造句，运用合适的方法进行转换，从而解决这些问题。

翻译方法纷繁众多，针对不同的语言对、翻译难点、文本类型等，中外学者提出了许多翻译方法。比如维奈和达贝尔内[2]（Vinay & Darbelnet, 1958）针对英法翻译提出了7种主要方法，范·道斯莱尔[3]（Van Doorslaer, 2007）在译学结构图中归纳了翻译的24种方法和技巧，张培基[4]针对英译汉翻译总结了15种常见方法。这些方法有的指代相同而名称不同，有的概念之间存在交叉。由于维奈和达贝尔内的模式提出时间较早，传播范围较广，因此，我们选择在其基础上进行一定的补充，并针对航空航天文学的翻译进行分析。综合来看，译者常采用以下方法。

---

[1] 熊兵. 翻译研究中的概念混淆：以"翻译策略"、"翻译方法"和"翻译技巧"为例[J]. 中国翻译，2014（3）：82-88.

[2] Vinay J, Darbelnet J. Comparative Stylistics of French and English: A Methodology for Translation [M]. Amsterdam / Philadelphia: John Benjamins, 1995.

[3] Van Doorslaer L. Risking Conceptual Maps[M]//Gambier Y, L. van Doorslaer. The Metalanguage of Translation Philadelphia: John Benjamins Pub. Company, 2007: 217-233.

[4] 张培基. 英汉翻译教程[M]. 上海：上海外语教育出版社，2018.

1. 异化策略下的翻译方法

- 借用(borrowing)
- 仿造(calque)
- 直译(literal translation)

2. 归化策略下的翻译方法

- 置换(transposition)
- 调适(modulation)
- 对等(equivalence)
- 显化(explication)
- 隐化(implicitation)
- 改编(adaptation)

接下来,我们将结合具体文本,阐述如何将这些技巧和方法应用于航空航天文学英汉翻译。

一、借　用

借用是所有翻译方法中最简单的一种[①],译者可以直接借用原语中的某些词汇,不进行翻译和转换。这种方法一般常用于以下两种情况:一是当目标语言中没有对应的词汇来表达原语言中的特定术语或文化现象时,译者会采用借用法从而确保准确性;二是当译者想要保留原语言中的某种特定表达时,会采用借用法从而达到特殊的表达效果。这种方法适用于那些在目标语言中已经被广泛接受或认可的外来词汇,或者是可以根据上下文或共同知识被目标读者理解的外来词汇。在航空航天文学中,许多人名、地名、机构名、专门概念和专业技术等都采用了借用法。

直接借用词形,也就是零翻译(zero translation)在英译汉中出现频率不高,一般用来处理那些在国际上已经通用或公认的专业术语或缩略语,或者是目标语言中已经被广泛使用的外来词汇。例如 NASA(The National Aeronautics and Space Administration,美国国家航空航天局)、VIP(Very Important Person,贵宾)、Wi-Fi(Wireless Fidelity,无线局域网)等。

除了保留词形外,一些词汇需要借用原语发音,确保翻译的准确度,例如:

例 6　英文原文:

In just one year, 1945, it had been the site of the Trinity explosion—the first nuclear detonation—and the launch of the highest-flying rocket known

---

① Vinay J, Darbelnet J. Comparative Stylistics of French and English: A Methodology for Translation [M]. Amsterdam / Philadelphia: John Benjamins, 1995.

to man.①

中文译文：

仅仅一年后的 1945 年，这里就成了<u>托立尼提</u>（Trinity，人类历史上首次核试验）的试验场，有史以来飞得最高的火箭也是在这里升空的。②

在本例中，划线部分 trinity 是核试验的代号，该词意为"三位一体"。但这一指称不具有强指向性，它还可以表示神学中的概念。由于目前并未公开该命名的具体来源，并未解释为什么叫作这一名称，因此，汉语译文保留了与原语相似的发音，从而确保指代的准确性，避免产生歧义。

例 7　英文原文：

So perhaps it is not surprising that all the oldest names for Mars have a peculiar weight on the tongue——<u>Nirgal, Mangala, Auqakuh, Harmakhis</u>——they sound as if they were even older than the ancient languages we find them in, as if they were fossil words from the Ice Age or before.③

中文译文：

这也难怪火星的古名在舌尖的发音都那么沉重——尼尔格、蒙加拉、安夸库，以及哈马契斯——好像比它们系出同源的古代语言更加古老，仿佛是冰川时代或是更早以前遗留下来的语言化石。④

例 7 划线部分均为火星的古名，这里是举例说明这些名称在发音上具有共性，因此译者根据原语保留发音特点，而非翻译其指称意义。

如果可以兼顾，译者还会在采用音译的同时也适当体现其意义，例如将 engine 译为"引擎"，既在发音上与原语相似，同时"引"在一定程度上也可以指示出这一表达的功能。

二、仿　造

仿造是指把原语言中的构成元素直译到目的语中，构成新词⑤。可以说，仿造也是一种特殊形式的借用，只不过借用的是词的意义而非形式。在航空航天文学中，一些专业术语和特殊指称的翻译常常会使用仿造法。

例 8　英文原文：

By 1506 Eurasia was coming into focus and there were glimmers of South America. By 1570 the Americas were coming into focus, but there was no

---

① Nathalia H. Rise of the Rocket Girls[M]. New York：Little, Brown and Company, 1976.
② 娜塔莉娅·霍尔特. 让火箭起飞的女孩[M]. 阳曦，译. 北京：九州出版社，2022.
③ Kim S R. Red Mars[M]. New York：Bantam Books, 1993.
④ 金·斯坦利·罗宾逊. 红火星[M]. 王凌霄，译. 重庆：重庆出版社，2016.
⑤ Vinay J, Darbelnet J. Comparative Stylistics of French and English：A Methodology for Translation[M]. Amsterdam / Philadelphia：John Benjamins, 1995.

sign of Australia. By 1744 Australia was coming into focus, but Antarctica was terra incognita.①

中文译文：

1506年，欧亚大陆的版图已然清晰，但南美洲仍在"朦胧"之中。1570年，美洲的轮廓已经清晰，但地图上仍然没有澳洲的丝毫踪迹。1744年，澳洲的版图也清晰了，但南极洲仍是未知领域。②

本例中，划线部分 terra incognita 源自拉丁语，terra 指"地球、土地"，incognita 指"未知的"，译文将各组成部分直译，并按照中文的习惯表达法表述，译为"未知领域"，运用了仿造的方法。

英汉属于两种不同的语系，因此有时很难保留原语文字的书面形式和结构。在这种情况下，仿造法能够尽可能地模仿原语的发音、形式、结构等，增强与目标语言和文化中的关联性，将信息处理成为目标语言读者更容易理解和接受的表达方式，可以增强译文的可读性，使其在目的语文化中得到更好的传播③，并逐渐被大众所熟知且接受，例如 artificial satellite（人造卫星）、outer space（外太空）、manned spacecraft（载人飞船）、light-years（光年）、space station（空间站）等。

三、直　译

直译也被称为字面翻译（word-for-word translation），指按照原语言的字面意思进行翻译，并保持词汇、句法和语序的对应。这也是在相同语系或文化语境的语言中，最普遍的一种翻译方法④。然而，由于英汉两种语言之间存在较大差异，很少能够保证译文与原文结构相同，完全对应。因此，直译只能做到最大限度地还原原文。

例9　英文原文：

It is a counting house in the wilderness—a place of shillings and pounds and land sales and trade, extraordinary successes and extraordinary failures.⑤

中文译文：

它是荒野中的财会室——这地方关乎先令、英镑、土地买卖、贸易，关乎极度成功以及极度失败。⑥

然而，英语和汉语中存在一些形式和结构相互对应的表达，但内涵却不相同。此

---

① Kip T. The Science of Interstellar[M]. New York：W. W. Norton & Company, 2014.
② 基普·索恩. 星际穿越[M]. 苟利军,王岚,李然,等译. 杭州：浙江人民出版社, 2015.
③ 樊林洲, 陈胜男. 英汉术语仿造翻译法诠索[J]. 中国科技术语, 2022(4)：53-60.
④ Vinay J, Darbelnet J. Comparative Stylistics of French and English：A Methodology for Translation[M]. Amsterdam / Philadelphia：John Benjamins, 1995.
⑤ Beryl M. West with the Night[M]. New York：North Point Press. 1983.
⑥ 柏瑞尔·马卡姆. 夜航西飞[M]. 陶立夏,译. 北京：人民文学出版社, 2016.

时,译者应当注意表达的具体意义,鉴别"假朋友"现象。

例 10 英文原文:

Although she loved her work, with Richard's promotion and subsequent addedincome, she was thinking about <u>starting a family</u>.①

中文译文:

虽然她热爱自己的工作,但既然理查德升了职又加了薪水,她开始考虑<u>生孩子了</u>。②

本例中,原文中为"start a family",而译文为"生孩子"。这是因为这一表述在中英文化中略有差异,英语中强调孩子是家庭的关键组成部分,因此 start a family 指的是夫妇决定要生育孩子。而中文中"组建家庭"的概念更为广泛,可以是结婚,不仅仅强调生育孩子。所以这里不能按照字面形式进行翻译,从而避免汉语读者产生误解。

在航空航天领域中,专业词汇与普通常见的名词词形相同,但实际含义也会有很大差异。比如 apron 在日常生活中指"围裙",而在航空领域中指的是"停机坪";trim 在日常生活中指"修剪,削减",而在航空领域中指的是"调整飞机负载以保持平衡和稳定,使飞机配平";taxi 常见的意思为"出租车",而在航空领域中指的是"飞行员驾驶飞机缓慢滑行"。因此,我们在运用直译法时,需要注意这些词汇在航空航天领域指代的内容是什么,并注意查证,从而确定其专业名称,例如:

例 11 英文原文:

As they <u>banked</u>, Dr. Floyd could see below him a maze of buildings, then a great airstrip, then a broad, dead-straight scar across the flat Florida landscape—the multiple rails of a giant launch-lug track. At its end, surrounded by <u>vehicles and gantries</u>, a spaceplane lay gleaming in a pool of light, being prepared for its leap to the stars.③

中文译文:

随着飞机<u>侧弯</u>,弗洛伊德博士可以看到下方迷宫般的建筑群,接着是一条大跑道,然后是一条又宽又直、横越佛罗里达平坦地面的疤痕——这是一条巨大的多轨发射道。跑道尽头,在各种<u>载具和支架</u>的环绕下,一艘宇宙飞船在一片灯光下闪闪发亮,正准备跃入星空。④

本例中,划线部分 bank 作为动词一般指的是"把钱存入银行",而在这一语境中,指的是"飞机转弯时倾斜飞行"。此外,vehicle 常指"车辆",gantry 常指"(路标、信号灯等)金属支架",结合上下文可以推断,这里指代的是辅助宇宙飞船升空的设

---

① Holt N. Rise of the Rocket Girls[M]. New York: Little, Brown and Company, 2016.
② 娜塔莉娅·霍尔特. 让火箭起飞的女孩[M]. 阳曦,译. 北京:九州出版社,2022.
③ Clarke C A. 2001: A Space Odyssey[M]. UK: Pearson Education Ltd & Penguin Books Ltd, 1968.
④ 阿瑟·克拉克. 2001:太空漫步[M]. 郝明义,译. 上海:上海文艺出版社,2019.

施,因此,译文将其分别处理为"载具"和"支架",避免产生歧义,让读者误解。

同时,由于作者通常会通过自己的亲身经历来呈现出真实的航空航天领域历史事件、事故和人物等,译者也需要注重细节,尽可能地保持内容的准确性和真实性。

例 12　英文原文:

It was in the army, while serving as an honorary colonel, that Tsien consulted for Operation Paperclip, which aimed to capture key Nazi scientists after the war before Russia could get hold of them.①

中文译文:

在军队里担任荣誉上校期间,钱学森曾是回形针行动的顾问,这个计划的目标是抢在苏联人的前面抓住纳粹的重要科学家。②

例 13　英文原文:

JPL launched the Mars Pathfinder, armed with airbags, in December 1996. The mood was tense; after all, it had been two decades since a successful Mars mission. The loss of the Mars Observer had extended the long lull in Martian exploration. Russia had similarly struggled in getting to the Red Planet.③

中文译文:

1996 年 12 月,JPL 发射了配备气囊的火星探路者号(Mars Pathfinder)飞船。实验室里的情绪十分紧张,归根结底,上一次成功的火星任务已经是 20 年前的事儿了。火星观察者号的失败无疑延长了火星探索的低迷期。俄罗斯的火星探索任务同样举步维艰。④

在例 12 和例 13 中,两个句子的原文均为"Russia",而在译文当中,前者译为"苏联",后者译为"俄罗斯",就是因为两处事件发生的时间不同,后者处于苏联解体之后,因此在译文中进行区分,做到真实准确,符合史实。

四、置　换

置换法是指在翻译过程中为了使译文符合目标语言的表达方式,对原语言中的语法范畴进行转换。置换可以是强制性的,也可以是选择性的。

英语和汉语中存在许多不对应的词类,例如英语中有关系代词、关系副词等,而汉语中没有;英语中使用动词需要考虑人称、数、时态、语态等较多限制,而汉语在使用动词时基本没有限制;英语中不同词性的功能和用法差异明显,而汉语中对词性并

---

① Holt N. Rise of the Rocket Girls[M]. New York: Little, Brown and Company, 2016.
② 娜塔莉娅·霍尔特. 让火箭起飞的女孩[M]. 阳曦,译. 北京:九州出版社,2022.
③ 同①.
④ 同②.

不过度强调。

　　对于航空航天文学的英汉翻译，我们需要基于原文的意义，根据汉语的表达习惯进行置换，从而使译文自然流畅。从词性方面，译者可以灵活选择。可以把名词转换为动词、代词、形容词、副词，把形容词转换为副词、名词，动词也可以转换为名词、形容词等。同时，随着词性发生变化，句子成分也会不同。主语可能变成宾语、表语、定语、状语，谓语可能变成主语、定语，宾语可能变成主语等。通过置换，词与句之间的关系进行调整，整个句式有时也会发生改变。

1. 名词的转换

例 14　英文原文：

A gentle pull on the throttle eased the motor to an effortless hum.①

中文译文：

轻轻牵引油门，让引擎发出放松的低鸣。②

本例将名词 pull 转换成了动词"牵引"。若不进行转换，在译文中依然保留名词，就会变成"对油门轻轻的拉动使引擎发出放松的低鸣"，读起来就会带有翻译腔，不够自然。

例 15　英文原文：

Until my wanderings were checked by the discovery that I was wearing out my boots.③

中文译文：

直到我发现靴子都磨烂了，才不再游荡。④

本例中的名词 wandering 转换成动词"游荡"，名词 discovery 转换成动词"发现"。同时，由于词性发生改变，信息在句子中的成分改变，句式也由被动变为主动。

例 16　英文原文：

She felt my socialism, felt my spirit in revolt against the accepted order, felt the impotent resentments that filled me with bitterness against all she held sacred.⑤

中文译文：

她感觉出我信仰社会主义，感受到我的灵魂深处反感现有制度，体会到我对她所信奉的神圣事物都充满怨恨，却又无能为力，只能暗自抱怨。⑥

---

①　Beryl M. West with the Night[M]. New York：North Point Press. 1983.
②　柏瑞尔·马卡姆. 夜航西飞[M]. 陶立夏，译. 北京：人民文学出版社，2016.
③　Herbert G W. In the Days of the Comet[M]. Auckland：The Floating Press, 2009.
④　赫伯特·乔治·威尔斯. 彗星来临[M]. 穆雷，译. 大连：大连理工大学出版社，2018.
⑤　同③。
⑥　同④。

本例中,将名词 socialism 转换为"信仰社会主义",名词 revolt 转换为动词"反感",读起来更加自然流畅,清晰易懂。

例 17　英文原文:

I noticed that, in spite of his spirit and his courage, his voice had grown thin and less certain of its strength.①

中文译文:

我留意到,尽管他意志坚强、勇敢无畏,但他的声音越来越微弱,越来越不确定自己的音量。②

本例中,将英语中描述性格的名词 spirit、courage 转换为汉语形容词"意志坚强""勇敢无畏",符合目的语的表达习惯,使译文更顺畅。

例 18　英文原文:

Here I had spent the afternoon, experimenting and practising with careful deliberation and grim persistence.③

中文译文:

我在这里待了一个下午,仔细从容、冷静固执地试验与练习。④

本例中,将英语中的名词词组转变为汉语副词,同样采用了置换的方法。

2. 形容词的转换

例 19　英文原文:

Beyond, the High Street with shops resumed again in good earnest and went on, and the lines of the steam-tramway that started out from before my feet, and were here shining and acutely visible with reflected skylight and here lost in a shadow, took up for one acute moment the greasy yellow irradiation of a newly lit gaslamp as they vanished round the bend.⑤

中文译文:

远处,商店林立的高街重新开张,继续热闹起来,蒸汽机车的行驶线路就从我脚下延伸开去,可以清晰地看见反射的天光在闪烁,然后消失在阴影里,有那么一会儿,一盏新点燃的煤气灯发出油腻腻的黄色光亮,再逐渐消失在拐弯处。⑥

本例中,在描述蒸汽机车时,运用形容词 shining 和 visible,但若在译文中照搬,译为"闪耀且清晰可见的",便使语言失去了韵味。译文将形容词转换为动词,读起来

---

① Beryl M. West with the Night[M]. New York: North Point Press. 1983.
② 柏瑞尔·马卡姆. 夜航西飞[M]. 陶立夏,译. 北京:人民文学出版社,2016.
③ Herbert G W. In the Days of the comet[M]. Auckland: The Floating Press, 2009.
④ 赫伯特·乔治·威尔斯. 彗星来临[M]. 穆雷,译. 大连:大连理工大学出版社,2018.
⑤ 同③。
⑥ 同④。

更加地道自然。

例 20　英文原文：

At first it had been an almost telescopic speck; it had brightened to the dimensions of the greatest star in the heavens; it had still grown, hour by hour, in its incredibly swift, its noiseless and inevitable rush upon our earth, until it had equaled and surpassed the moon.①

中文译文：

起初，它看起来就像望远镜中的微点；然后慢慢变亮，亮到像是天空中最大的星星一般；继而持续增大，以令人难以置信的速度迅速增大。它无声无息、不可避免地冲向地球，直到与月球同样大小，最后超过月球。②

本例中，在描述彗星接近地球时，将形容词 swift、noiseless、inevitable 分别转换成了副词"迅速""无声无息""不可避免地"，符合汉语的表达习惯。

3. 介词的转换

例 21　英文原文：

I had walked out with it four or five miles across a patch of moorland and down to a secluded little coppice full of blue-bells, halfway along the high-road between Leet and Stafford.③

中文译文：

（我）握着枪走出四五英里，穿过一小片沼泽地，走进一片偏僻的灌木丛，里面长满风铃草。这里位于利特与斯塔福德之间的公路的一半处。④

本例中，在介绍主人公行动时，原文运用了三个介词，但在译文中将其都转换为动词，明晰主人公的动作，同时具有"移步换景"的效果，符合汉语读者的阅读习惯。

综合而言，置换法具有较强的灵活性，我们不应该拘泥于原语文本的词性，而要学会灵活变通。不过从整体上说，英汉翻译的词性转换表现出一定的规律，英语名词转化为汉语动词的情况较多。这是因为英语是静态语言，使用动词较少，而倾向于使用名词、介词，而汉语是动态语言，使用动词较多。

五、调　适

当采用直译或者置换法时，译文在目的语中可能是"不恰当、不地道，甚至糟糕

---

① Herbert G W. In the Days of the Comet[M]. Auckland: The Floating Press, 2009.
② 赫伯特·乔治·威尔斯. 彗星来临[M]. 穆雷, 译. 大连：大连理工大学出版社, 2018.
③ 同①。
④ 同②。

的"①,这时我们就可以采用调适法。与句法层面的置换不同,"调适"强调语义层面,针对的是语法正确,但在目的语中听起来不自然的表达方式。反应的是说不同语言的人在看待事物和现象时,在视角上体现出的细微差别。这种方法对译者在语言能力方面有更高的要求,因此也被视为"优秀译者的试金石"。②

由于英语母语者与汉语母语者在思维方式、语言习惯方面存在很大的差异,调适法在英汉翻译中较为常见。在航空航天文学英译汉中,译者进行的调适主要有以下几类:

1. 抽象转为具象

例 22 英文原文:

I have seen a herd of buffalo <u>invade the pastures</u> under the occasional thorn tree groves and, now and then, the whimsically fashioned <u>figure</u> of a plodding rhino has <u>moved</u> along the horizon like a grey boulder come to life and adventure bound.③

中文译文:

我曾看见一群水牛在偶尔出现的棘树下<u>吃草</u>,突然,模样怪异的犀牛蹒跚着<u>走过</u>地平线,仿佛一块灰色的巨石拥有了生命,来到野外。④

本例中,invade the pastures 从字面意义来看翻译为水牛"入侵牧场",如果照搬,读起来可能就会觉得不知所云。因此在译文中转换为更加易懂的具体动作"吃草"。同样,对于 figure、move,译者也采用了更加具象的表达。

例 23 英文原文:

I had brought an old kite-frame of cane with me, that folded and unfolded, and each shot-hole I made I marked and numbered to compare with my other <u>endeavors</u>.⑤

中文译文:

我带了一个藤制的风筝骨架,打开合上,合上又打开,每次打出弹孔,我就会做出标记,数出环数,再和其他<u>环弹孔</u>作比较。⑥

本例中 endeavor 意为"努力",结合上下文可知,其指代的是主人公的射击结果。译文将原文的抽象表述转变为具象的"弹孔",使译文表述更加清晰易懂。

---

① Vinay J, Darbelnet J. Comparative Stylistics of French and English: A Methodology for Translation[M]. Amsterdam/Philadelphia: John Benjamins, 1995.
② 同①。
③ Beryl M. West with the Night[M]. New York: North Point Press. 1983.
④ 柏瑞尔·马卡姆. 夜航西飞[M]. 陶立夏,译. 北京: 人民文学出版社, 2016.
⑤ Herbert G W. In the Days of the Comet[M]. Auckland: The Floating Press, 2009.
⑥ 赫伯特·乔治·威尔斯. 彗星来临[M]. 穆雷,译. 大连: 大连理工大学出版社, 2018.

例 24　英文原文：

A blur of faces surrounded her, all gawking at the scene, unsure of exactly what they were witnessing.①

中文译文：

周围的学生们都目瞪口呆地望着眼前的场景,不知道出了什么事。②

本例中,如果不进行解释,译文读起来就会十分奇怪。将 faces 转换为表意更加具体清楚的"学生们",使指代更加明晰,易于读者理解。

2. 被动转为主动

在较为严谨的科技类文本翻译中,被动句转变成主动句尤为明显。相较而言,汉语使用被动句的频率较低,因此,译者在英汉翻译时往往会采用多种手段避免被动,将其转化为主动句,或者隐化被动关系,例如：

例 25　英文原文：

There's not much to be seen in light like that—some dark upturned faces impassive and patient, half-raised arms beckoning, the shadow of a dog slouching between the flares.③

中文译文：

那样的光线中什么都看不清楚：几张仰望的黝黑脸庞,神色冷漠而坚忍；几条半举着的手臂,做着召唤的姿势；有条狗懒洋洋地穿行在火光中……④

英语原文 there's not much to be seen 意为"没有什么被看到的",译文将其转换为主动语态："什么都看不清楚",读起来更加流畅自然。

例 26　英文原文：

For the trampoline, the warping of space is produced by the rock's weight. Similarly, one might suspect, the black hole's space warp produced by the singularity at its center. Not so. In fact, the hole's space is warped by the enormous energy of its warping.⑤

中文译文：

蹦床的弯曲是由石头的重量产生的。你可能会猜想,类似地,黑洞的空间弯曲也来自奇点。但并非如此。事实上,黑洞的空间弯曲是由引起空间弯曲的巨大能量造成的。⑥

---

① Holt N. Rise of the Rocket Girls[M]. New York: Little, Brown and Company, 2016.
② 娜塔莉娅·霍尔特. 让火箭起飞的女孩[M]. 阳曦,译. 北京：九州出版社,2022.
③ Beryl M. West with the Night[M]. New York: North Point Press. 1983.
④ 柏瑞尔·马卡姆. 夜航西飞[M]. 陶立夏,译. 北京：人民文学出版社,2016.
⑤ Kip T. The Science of Interstellar[M]. New York: W. W. Norton & Company, 2014.
⑥ 基普·索恩. 星际穿越[M]. 苟利军,王岚,李然,等译. 杭州：浙江人民出版社,2015.

本例中,原文采用被动表达,译文将 be produced by 译为"由……产生",将 be warped by 译为"由……造成",均隐化了表示被动的结构。

例 27　英文原文:

The three members of the survey team, who would not be needed until the ship entered her final orbit around Saturn, would sleep through the entire outward flight. Tons of food and other expendables would thus be saved; almost as important, the team would be fresh and alert, and not fatigued by the ten-month voyage, when they went into action.

中文译文:

到宇宙飞船最后进入环绕土星的轨道之前,这整段向外飞行的过程中,探勘队有三位成员无须参与,可以一直沉睡。这样可以省下大量食物及其他消耗品。还有一点很重要的是,等他们醒来进入工作岗位的时候,可以精神抖擞,不会有航行十个月的疲惫。

本例中,原文 would not be needed 译为"无需参与",be saved 译为"省下",would be not fatigued by 译为"不会有……疲惫",将句子中所有被动都转化成了主动表达。

3. 物称转为人称

英语中常用物称进行表达,即不使用人称作为主语进行叙述,描述客观概念或事物对人产生什么结果,而中文则相反,习惯描述某人如何影响客观事物①。在这种情况下,就需要译者对人称和物称进行转换。

例 28　英文原文:

The Suicide Squad's amplified destruction had gotten them kicked off the Caltech campus.②

中文译文:

慑于自杀小队越来越强的破坏力,加州理工将他们赶出了校园。③

本例中,若保留原本的句式结构,就变成"自杀小队越来越强的破坏力让他们被赶出加州理工校园",读起来远没有译文流畅自然。

例 29　英文原文:

How hard the pendulum swung translated into how high a rocket might one day fly.④

---

① 连淑能. 英汉对比研究[M]. 北京:高等教育出版社,2010.
② Holt N. Rise of the Rocket Girls[M]. New York: Little, Brown and Company, 2016.
③ 娜塔莉娅·霍尔特. 让火箭起飞的女孩[M]. 阳曦,译. 北京:九州出版社,2022.
④ 同②.

中文译文：

只需要测量单摆的振幅，他们就能算出火箭在发射试验中可能达到的高度。①

本例中，原文采用物称结构，使句子读起来比较客观。译文突出了主体视角，其中的"测量""算出"都是人施加的动作，更符合中文的表达习惯。

例30　英文原文：

Bergner closed his eyes and let a tremor of pain shake his body under the flimsy blanket.②

中文译文：

伯格纳闭上眼睛，薄毯下的身躯因为痛楚而颤动。③

本例中，原文为"疼痛使他的身体颤抖"，译文将其转换为"他因疼痛而颤抖"，主体对象由物转为人，使表达更加自然。

4. 调整对象顺序

我们知道，英语重形合，汉语重意合。英语句子中，词汇和成分之间可以通过语言形式手段连接起来，因此通常以主谓结构为主干，凸显主要描述对象，同时借助介词、连词、关系代词、关系副词等把次要成分纳入句子内容。汉语则不然，主要通过连用动词来体现时间先后、空间顺序、逻辑顺序，先讲背景后讲事件，由远及近，由因及果。因此，在英译汉中，译者往往会对句子进行分析，调整描述对象的顺序，按照汉语母语者的认知习惯重新组合。

例31　英文原文：

He had commandeered an old opera-glass from his uncle who farmed at Leet over the moors, he had bought a cheap paper planisphere and *Whitaker's Almanac*, and for a time day and moonlight were mere blank interruptions to the one satisfactory reality in his life—star-gazing.④

中文译文：

他的叔叔在荒野那边的利特经营农场，那架看戏用的小望远镜正是他从他叔叔那里要来的。此外，他还买了一张廉价的纸本星座图和一本《惠特克历书》。在那段时间里，观星是他生命中唯一满足的现实，而日光高照或月光太亮时会扰乱观星。⑤

本例中，原文整句都是围绕 he，利用从句、分词、连词描述次要成分。若按照原文的顺序和结构进行翻译，译文就会过长且复杂。中文习惯先介绍背景，建立起与主体事物的联系，然后引出主要内容，再由主要内容延伸至具体细节。因此，译者将描

---

① 娜塔莉娅·霍尔特. 让火箭起飞的女孩[M]. 阳曦, 译. 北京：九州出版社, 2022.
② Beryl M. West with the Night[M]. New York: North Point Press. 1983.
③ 柏瑞尔·马卡姆. 夜航西飞[M]. 陶立夏, 译. 北京：人民文学出版社, 2016.
④ Herbert G W. In the Days of the Comet[M]. Auckland: The Floating Press, 2009.
⑤ 赫伯特·乔治·威尔斯. 彗星来临[M]. 穆雷, 译. 大连：大连理工大学出版社, 2018.

述对象进行调整,先交代了背景信息"叔叔"与"小望远镜",然后才介绍主人公的相关事件,符合汉语读者的思维逻辑。

例 32　英文原文：

My search came up with an old black-and-white photo of her, blond bouffant hair curling at her shoulders, a timid smile as she held up an astronomy award for her asteroid discoveries.①

中文译文：

我还搜到了一张赫林的黑白照片,她捧着一面天文学奖牌,嘴角噙着一抹羞涩的微笑,金色的卷发蓬松地堆在肩头。②

本例中,原文对于描述对象的顺序安排较为灵活,依次提到了照片、卷发、微笑和奖牌。译文对这些指称的顺序进行了调换,更符合汉语读者的认知过程,即先整体后细节。

5. 反向表述

反向表述指的是在翻译过程中,将原文中表达的意思从相反的方向重新进行表述,以达到更好的传达效果。例如,在原文中,可能会使用否定词、强调词等语言手段来表达某种情感或意义。在翻译时,译者可能会将这些手段转化为肯定词、弱化词等。这种方法可以帮助译者更好地传达原文的意义,使翻译结果更加贴近目的语的表达方式,减轻读者阅读的难度。

例 33　英文原文：

She had been awake for more than sixteen hours but felt no fatigue.③

中文译文：

她已经有 16 个小时没睡觉了,但她丝毫不觉得疲惫。④

例 34　英文原文：

(The house was) not a remarkable house by any means—it was about thirty years old, squattish, squarish, made of brick, and had four windows set in the front of a size and proportion which more or less exactly failed to please the eye.⑤

中文译文：

这幢屋子不管从任何意义上说都平平常常,房龄快三十年了,矮胖短粗,方头方

---

① Holt N. Rise of the Rocket Girls[M]. New York: Little, Brown and Company, 2016.
② 娜塔莉娅·霍尔特. 让火箭起飞的女孩[M]. 阳曦,译. 北京:九州出版社,2022.
③ 同①。
④ 同②。
⑤ Douglas A. The Hitchhiker's Guide to the Galaxy[M]. New York: Ballantine Books, 2004.

脑,砖木结构,正面的四扇窗户不管是尺寸还是比例都或多或少地让人看了不舒服。①

此外,在描述方位时,译者也会根据目的语的语言结构和习惯进行调整,将"A在B的前方"说成"B在A的后方",当然这种情况下描述对象的顺序会发生一定的变动,例如:

例 35　英文原文:

The object before which the spacesuited man was posing was a vertical slab of jet-black material, about ten feet high and five feet wide; it reminded Floyd, somewhat ominously, of a giant tombstone.②

中文译文:

穿着航天服的人后方,直立着一块漆黑质地的板子,大约有十英尺高、五英尺宽。弗洛伊德多少有点不吉利地联想到一块巨大的墓碑。③

例 36　英文原文:

The general effect, as one came down the hill, was of a dark compressed life beneath a very high and wide and luminous evening sky, against which these pit-wheels rose.④

中文译文:

从山上走下来,整体的印象是一幅灰暗压抑的生活图景,上面是高深广袤、闪闪发亮的夜空,煤矿的那些大轮子就映在夜色之中。⑤

六、对　等

对等指在目标语言中寻找与原语言具有相同或相近意义但形式不同的表达方式进行翻译,这一方法尤其适合处理一些固定搭配、惯用法、俗语、俚语、谚语的翻译。

例 37　英文原文:

Even to an untrained eye, it was obvious that something peculiar had happened to the Moon's magnetic field in this region.⑥

中文译文:

就算是外行人,也看得出月球这个地区的磁场发生了什么很特别的事情。⑦

本例中,untrained 意思是"未经过训练的",比如在艺术领域,一位没有经过专业

---

① 道格拉斯·亚当斯. 银河系搭车客指南[M]. 姚向辉,译. 上海:上海译文出版社,2011.
② Clarke C A. 2001: A Space Odyssey[M]. UK: Pearson Education Ltd & Penguin Books Ltd, 1968.
③ 阿瑟·克拉克. 2001:太空漫步[M]. 郝明义,译. 上海:上海文艺出版社,2019.
④ Herbert G W. In the Days of the Comet[M]. Auckland: The Floating Press, 2009.
⑤ 赫伯特·乔治·威尔斯. 彗星来临[M]. 穆雷,译. 大连:大连理工大学出版社,2018.
⑥ 同②。
⑦ 同③。

训练的人可能无法识别出名画与普通画作之间的区别。所以,这一用法常用来描述缺乏特定技能或专业知识的人。汉语中的"外行人""门外汉"意义与此相似。

例38 英文原文:

It was foolhardy to expect success right off the bat. JPL was used to hard-won success born from repeated failure.[1]

中文译文:

指望一蹴而就无异于天方夜谭,JPL早就习惯了历经磨难才能修成正果。[2]

本例中,right off the bat是一个习语,意为"毫不犹豫地、立即地"。这一表述源自19世纪末期的棒球游戏,因为当球棒击中棒球,棒球就会立即以非常快的速度弹出去。原文的意思是说,期待立即实现成功并不现实,译者采用了中文中形式不同但意义相近的表达"一蹴而就",易于理解。

例39 英文原文:

Her professor was impressed with her moxie and soon came to realize that Sue was more than a pretty face in his otherwise all-male classroom.[3]

中文译文:

这个女孩的勇气和精力让教授深感诧异,很快他就意识到,在这个全是男人的教室里,苏绝不仅仅是个花瓶。[4]

本例中,a pretty face源自be just another pretty face这一表达,用于形容只是外表好看而缺乏实质能力和贡献的女性,这一概念与汉语中的"花瓶"对应,转换后使读者更易于理解。

七、显 化

显化是指对于原语中暗含但是可以从上下文中推导出的信息,译者会在译文中将其明示出来。这种方法与我们常说的增译技巧密切相关,但也存在一定的差异。比如一些学者就认为显化是翻译本身固有的特征,无论是何种语言,由于译者通常会在译文中进行解释和补充,降低读者阅读难度,因此一般要比原文更长,语义更清楚,逻辑衔接更紧密[5]。

在航空航天文学的英汉翻译中,显化方法主要有以下几种:

---

[1] Holt N. Rise of the Rocket Girls[M]. New York: Little, Brown and Company, 2016.
[2] 娜塔莉娅·霍尔特. 让火箭起飞的女孩[M]. 阳曦,译. 北京:九州出版社, 2022.
[3] 同[1].
[4] 同[2].
[5] Séguinot C. Pragmatics and Explicitation Hypothesis[M]. TTR: Traduction, Terminologie, Redaction, 1988(2): 106-113.

1. 明晰指代

例 40　英文原文：

In the late 1930s, this country was awakening to a sense of its own unpreparedness for war, and for several years Lockheed had been at work secretly developing a new fighter for the Army Air Corps.①

中文译文：

20 世纪 30 年代后期,美国才开始警觉到对战争毫无准备。实际上,洛克希德公司已花了好几年时间为陆军航空队研制新型战斗机。②

英语中多用指示代词进行照应,而汉语中较少使用这些词,一般就会解释出指代的具体内容,以方便读者理解,如在本例中,this country 指代的就是"美国"。

例 41　英文原文：

And then the world began to change. In four great waves, with two hundred thousand years between their crests, the Ice Ages swept by, leaving their mark on all the globe. Outside the tropics, the glaciers slew those who had prematurely left their ancestral home; and everywhere they winnowed out the creatures who could not adapt.

中文译文：

然后,世界也开始改变了。四波大冰河期横扫而过,每一波高峰间隔二十万年,在地球到处都留下了标记。热带以外的地方,冰河消灭了贸然离开祖居地的动物,所到之处,没法适应的生物,就一一遭到淘汰。

原文中采用从句,只对 those 进行描述,其指代的具体内容并未指出,译文结合语境将其译为"动物",相比原文表达更加清楚,更容易读懂。

2. 明晰语义

例 42　英文原文：

A man can be riddled with malaria for years on end, with its chills and its fevers and its nightmares, but, if one day he sees that the water from his kidneys is black, he knows he will not leave that place again, wherever he is, or wherever he hoped to be.③

---

① Johnson C, Maggie S. Kelly. More Than My Share of It All[M]. Washington D. C.: Smithsonian Books, 1985.

② 凯利·约翰逊,玛吉·史密斯. 我怎样设计飞机:美国飞机设计师凯利·约翰逊自传[M]. 程不时,等译. 北京:航空工业出版社,1990.

③ Beryl M. West with the Night[M]. New York: North Point Press. 1983.

中文译文：

患疟疾的人可能经历数年的折磨才去世,承受着那些寒冷、高烧和噩梦。但是,假如某天,他发现自己的尿液变成了黑色,他就知道自己再无可能离开那个地方了,无论他身处何地,也无论他想去往何方。①

本例中,原文 water from his kidney 可直译为"他肾脏里的水",属于一种委婉用法,但若保留在译文中,就显得比较怪异,译文将其译为"尿液",更加清晰易懂。

例43 英文原文:

No fire was laid, only a few scraps of torn paper and the bowl of a broken corn-cob pipe were visible behind the bars, and in the corner and rather thrust away was an angular japanned coal-box with a damaged hinge.②

中文译文:

壁炉里没有生火,在炉栅后只能看见一些碎纸片和一碗掰断的玉米芯,角落更深处有一只棱角分明、刷了亮漆的煤箱,上面吊着一只破合页。③

本例将 broken corn-cob 译为"掰断的玉米芯",将 japanned 译为"刷了亮漆的",将 with 译为"吊着",对物品的状态进行更加明晰、具体的描述。让读者更容易理解原文的含义,并且更容易想象出原文中描述的场景。

例44 英文原文:

People tend to think astronauts have the courage of a superhero—or maybe the emotional range of a robot.④

中文译文:

人们往往认为宇航员有着超级英雄的胆略——或者是像机器人那样毫无个人情感。⑤

本例中,英语原文为 the emotional range of a robot,意思是说宇航员的情绪变动范围和机器人一样,译文中进一步进行阐释,将其中隐含的"毫无个人情感"解释出来。

例45 英文原文:

Mr. Gabbitas, the curate of all work, lodged on our ground floor, and upstairs there was an old lady, Miss Holroyd, who painted flowers on china and maintained her blind sister in an adjacent room; my mother and I lived in the basement and slept in the attics.

---

① 柏瑞尔·马卡姆. 夜航西飞[M]. 陶立夏,译. 北京:人民文学出版社,2016.
② Beryl M. West with the Night[M]. New York: North Point Press. 1983.
③ 同①.
④ Chris H. An Astronaut's Guide to Life on Earth[M]. Toronto: Random House Canada, 2013.
⑤ 哈德菲尔德. 宇航员地球生活指南[M]. 徐彬,译. 长沙:湖南科学技术出版社,2017.

中文译文：

加布比塔先生是一位全权代理牧师，寄宿在我们这幢房屋的一楼。他楼上是一位老姑娘——霍尔罗伊德小姐，以在瓷器上画花朵图案为生。住在隔壁房间的妹妹双目失明，靠她养着。我和母亲住在地下室，睡在阁楼上。

本例中，原文为 our ground floor，译文为"我们这幢房屋的一楼"，原文为 painted flowers on china，译文为"以在瓷器上画花朵图案为生"，通过解释细节，使语义更加清楚明白。

3. 弥补知识空缺

对于原文当中涉及特别知识背景的信息，难以在译文中简要表达出来，在这种情况下，译者有时会选择增加注释（文内注释或脚注），以方便读者理解。然而，相较于其他显化方法，增加注释会割裂文本，影响阅读的流畅性，因此译者一般都会尽可能在译文中巧妙地进行弥补，降低单独加注的频率。

例 46　英文原文：

In that period, I even tried to understand Einstein. Only 12 people in the whole world were supposed to be able to do so; I wanted to be the 13th![①]

中文译文：

在那段时间，我甚至还曾努力去弄懂爱因斯坦的相对论。假如全世界只有 12 个人能够懂得的话，我倒愿成为第 13 个人。（译者注：爱因斯坦发表《相对论》时，他说全世界只有 12 个人能懂得他提出的理论。[②]）

本例中，若读者不了解爱因斯坦曾经说过的这段内容，可能会觉得读起来跳跃，缺乏连贯性。

八、隐　化

与"显化"相反，"隐化"是将原文语境或情景的显性信息在译语中隐含起来的一种翻译方法。在英汉翻译中，一些在原语文化中存在，而在目的语文化中缺失的意象会被压缩。同时，若一些信息与译语文化的价值观念冲突，译者也会适当进行隐化，甚至删除。

例 47　英文原文：

He was dressed in his Sunday clothes, a sort of brownish tweeds, but the waistcoat was unbuttoned for greater comfort in his slumbers.[③]

---

①　Johnson C, Maggie S. Kelly. More Than My Share of It All[M]. Washington D. C. : Smithsonian Books, 1985.

②　凯利·约翰逊，玛吉·史密斯. 我怎样设计飞机：美国飞机设计师凯利·约翰逊自传[M]. 程不时，等译. 北京：航空工业出版社，1990.

③　Herbert G W. In the Days of the Comet[M]. Auckland：The Floating Press, 2009.

中文译文：

他身着盛装，那是一套棕色粗花呢衣服，但马甲敞着未扣，以便打盹时更舒适。①

本例中，Sunday clothes 与西方宗教文化有关。西方人有在周日时会去教堂做礼拜的习惯，即使是再穷的人也会穿上自己最好的衣服。中文中没有类似的表达，因此只能舍弃意象，译为"盛装"。

例 48　英文原文：

It also includes an oxidizer, an element such as oxygen that is able to accept an electron, thus setting in motion a powerful <u>oxidation-reduction reaction</u>, <u>often called a redox reaction</u>.②

中文译文：

它的成分还包括氧化剂，这种物质会像氧气一样吸收电子，引发剧烈的氧化还原反应。③

本例中，原文在提到化学反应的名称后，还介绍了它的简称，但是在中文中，这一反应只有一种固定的名称，因此后半部分内容被译者隐去。

例 49　英文原文：

Business goes on, banks flourish, automobiles purr importantly up and down Government Road, and <u>shop-girls</u> and clerks think, act, and live about as they do in any modern settlement of thirty-odd thousand in any country anywhere.④

中文译文：

生意在继续，银行蒸蒸日上，汽车在政府大道上煞有介事地来去，营业员们思考、行动、生活，他们在其他国家任何一个拥有三万多人口的现代城市里也会做同样的事。⑤

本例中描述的内容以 20 世纪上半叶为背景，shop girls 就是当时极具特色的一类群体——导购女郎。该职业均由女性承担，所以原文将其与 clerk 进行区分。译文中若按部就班地将其译为"导购女郎和售货员"，反而让人觉得累赘，难以理解，因此译文对其进行整合，只译其身份，而隐化性别因素。

例 50　英文原文：

Though birth control was cheap, reliable, and endorsed by all the main religions, it had come too late; the population of the world was now six

---

① 赫伯特·乔治·威尔斯. 彗星来临[M]. 穆雷, 译. 大连：大连理工大学出版社, 2018.
② Holt N. Rise of the Rocket Girls[M]. New York: Little, Brown and Company, 2016.
③ 娜塔莉娅·霍尔特. 让火箭起飞的女孩[M]. 阳曦, 译. 北京：九州出版社, 2022.
④ Beryl M. West with the Night[M]. New York: North Point Press. 1983.
⑤ 柏瑞尔·马卡姆. 夜航西飞[M]. 陶立夏, 译. 北京：人民文学出版社, 2016.

billion—a third of them in the Chinese Empire.①

中文译文：

虽然节育方法便宜又可靠，并且由各大宗教所支持，但还是来得太晚，全世界人口已经多达六十亿——其中三分之一在东方国家。②

例 51 英文原文：

And now, for their own inscrutable reasons, the Chinese were offering to the smallest have-not nations a complete nuclear capability of fifty warheads and delivery systems. The cost was under $200,000,000, and easy terms could be arranged.③

中文译文：

现在，基于一些高深莫测的动机，某些国家正在向一些贫穷小国家提供全套的配备：五十颗弹头外带火箭发射系统。开价不到两亿美元，而且条件好谈。④

例 50 和例 51 中，译者将原文中的"the Chinese Empire"和"the Chinese"分别翻译为"东方国家"和"某些国家"，对于原文中与目的语意识形态、受众认知不相符的内容进行了隐化处理。

九、改 编

改编是指为了适应译入语的社会文化背景，从而对原语中的信息进行改动。这种方法不拘泥于原语和目的语之间在概念上的对等，而强调情景对等。例如英语短语提到"板球"，这是英格兰人最喜欢的运动，那么在翻译成法语时，就可以用在法国地位相似的"环法自行车赛"来代替。改编可以说是"翻译的极限"⑤。

例 52 英文原文：

She confirmed her calculations before marking down the updated position on the graph paper. "She made it!" she said triumphantly, twisting around in her seat to see the reaction.⑥

中文译文：

她检查了自己刚才的计算，然后将卫星最新的位置标在了坐标纸上。"我们成功了！"她一边骄傲地宣布，一边转头去看大家的反应。⑦

---

① Clarke C A. 2001：A Space Odyssey[M]. UK：Pearson Education Ltd & Penguin Books Ltd, 1968.
② 阿瑟·克拉克. 2001：太空漫步[M]. 郝明义，译. 上海：上海文艺出版社，2019.
③ 同①.
④ 同②.
⑤ Vinay J, J Darbelnet. Comparative Stylistics of French and English：A Methodology for Translation [M]. Amsterdam / Philadelphia：John Benjamins, 1995.
⑥ Holt N. Rise of the Rocket Girls[M]. New York：Little, Brown and Company, 2016.
⑦ 娜塔莉娅·霍尔特. 让火箭起飞的女孩[M]. 阳曦，译. 北京：九州出版社，2022.

本例在描述卫星升空的结果时,原文说话人用的是 she made it,指卫星成功升空了,而译文中为"我们成功了",强调的是航空工作人员的努力结果。这反映出中西方思维方式上的差异,东方更加强调集体意识。虽然从语义上来看,似乎改变了原文内容,但从语用效果来看,这一处更加符合中文读者的认知和常识。

例 53 英文原文:

For five years, Tsien and his family were held under house arrest before being deported to China. ①

中文译文:

经历了 5 年的软禁,钱学森一家最终被遣送回了中国。②

本例中,原文的用词为 deport,指的是将(违法或者没有合法居留权的人)驱逐出境,而译文选用了"遣送"一词。其实"驱逐"与"遣送"在法律指称层面差别很小,但两个词在语言色彩上存在一定的差异。"驱逐"具有一定的负面含义,而"遣送"相比更加中性,没有明显的负面含义。钱学森是中国航天事业的奠基人,是为国家做出杰出贡献的科学家,因此对于中国读者来说,此处选择负面含义较弱的词汇更加合适。在描述相同历史事件时,不同国家的作者可能会选择不同的表达方式和评价标准,译者需要根据具体情况进行合理的选择。

总体来说,翻译方法是综合的,一句话中有时会灵活结合多种翻译方法。例如:

例 54 英文原文:

Perhaps after every other source had been tapped I might supplement with a few shillings frankly begged from her. ③

中文译文:

或许东拼西凑以后钱还不够的话,我可以向母亲坦白,乞求她给我几个先令。④

本例中,tap 作为动词表示"从某处索取钱",原文采用被动语态,译文将其转化为主动语态,采用了调适法。原文中只是提到了借钱这一行为,而译文结合语境对语义进行了显化,补充了"还凑不够的话",更方便读者理解。原文为 a few shillings frankly begged from her,即"(通过)坦白向她祈求得到的先令",对词性也进行了转换,运用了置换法。

例 55 英文原文:

When the electrodes had been attached to his forehead, and the sleep-generator had started to pulse, he had seen a brief display of kaleidoscopic patterns and drifting stars. Then they had faded, and darkness had engulfed him.

---

① Holt N. Rise of the Rocket Girls[M]. New York: Little, Brown and Company, 2016.
② 娜塔莉娅·霍尔特. 让火箭起飞的女孩[M]. 阳曦,译. 北京:九州出版社,2022.
③ Herbert G W. In the Days of the Comet[M]. Auckland: The Floating Press, 2009.
④ 赫伯特·乔治·威尔斯. 彗星来临[M]. 穆雷,译. 大连:大连理工大学出版社,2018.

中文译文：

当他的额头贴上了电极,睡眠产生器启动之后,他曾经短暂地看到一阵万花筒似的图案,以及漂流的星星。然后这些影像隐退,他进入无际的黑暗。

本例中,原文的 be attached to 被译为"贴上了",化被动为主动。原文为 display,意为"陈列,展览",在译文中,这一表述被隐去。原文为 darkness had engulfed him,即"黑暗吞没了他",译文为"他进入无际的黑暗",对陈述视角进行了转换。

练 习 题

一、基础练习：请翻译以下句子

1. This convoluted critical orbit is a close analog of the trajectories of temporarily trapped light rays inside Gargantua's shell of fire. ①

2. Like those light rays, the Endurance is temporarily trapped when on its critical orbit. ②

3. Unlike the light rays, the Endurance has a control system and rockets, so its launch off the critical orbit is in Brand's and Case's hands. ③

4. And because of the orbit's convoluted three-dimensional structure, the launch can be in any direction they wish. ④

二、拓展练习：请翻译以下段落

1. The Endurance heads toward Gargantua with a certain amount of energy, which like its angular momentum remains constant along its trajectory. This energy consists of three parts: the Endurance's gravitational energy, which gets more and more negative as the Endurance plunges toward Gargantua; its centrifugal energy (its energy of circumferential motion around Gargantua), which increases as the Endurance plunges because the circumferential motion is speeding up; and its radial kinetic energy (its energy of motion toward Gargantua). ⑤

2. Our discussion of Gargantua's environs has taken us from the physics of planets (tidal deformation, tsunamis, tidal bores,), through Gargantua's vibrations and the search for organic signs of life, to engineering issues (the Endurance's robust

---

① Kip T. The Science of Interstellar[M]. New York: W. W. Norton & Company, 2014.
② 同①。
③ 同①。
④ 同①。
⑤ 同①。

design and its damaging explosion). As much as I enjoy these topics—and I've done research or textbook writing on most of them— they are not my greatest passion. My passion is extreme physics; physics at the edge of human knowledge and just beyond. That's where I take us next. ①

---

① Kip T. The Science of Interstellar[M]. New York: W. W. Norton & Company, 2014.

第四章　航空航天文学翻译的策略与方法：汉译英

第一节　航空航天文学翻译汉译英的策略

在第三章中，我们对航空航天文学翻译的英译汉策略与方法进行了介绍，本章将对其汉译英策略与方法进行介绍。基本而言，汉译英与英译汉是互逆的，因此我们依然按照前文框架展开阐述，但对各策略和方法的具体概念在后文中将不再赘述。由于两种文化之间存在审美错位，加上文化交流不平衡等因素，英语和汉语之间的互译也体现出了一定的不对称性，因此依然需要分别进行研究①。

译者在选择翻译策略时大致有以下两种倾向：

一、异　化

例1　中文原文：
我心如刀绞地想象着妈妈最后的时刻，她同没能撤出的一万八千人一起，看着岩浆涌进市中心广场。②

英文译文：
Imagining my mother's final moments felt like twisting a knife in my heart. She, along with 18,000 others that could not be pulled out, must have seen the magma surge into the central plaza. (Holger Nahm 译)

本例中，"心如刀绞"是具有中国文化特色的成语，指内心痛苦，仿佛刀割。译文保留了这一意象，传神地体现出人物内心的痛苦情绪，同时使译文具有较强的表现力和异域特色。

二、归　化

例2　中文原文：
文章的观点可能不合一些人的胃口，但不要扣帽子，关键要看作者的长远思考。一些同志现在是一叶障目，有大环境的原因，也有很多人是自以为是。③

---

① 王建国. 英汉翻译学：基础理论与实践[M]. 北京：中译出版社，2020.
② 刘慈欣. 流浪地球[M]. 北京：中国科学技术出版社，2022.
③ 刘慈欣. 三体[M]. 重庆：重庆出版社，2008.

英文译文：

The article's views will not be to the liking of some, but let's not rush to label the author. The key is to appreciate the author's long-term thinking. Some comrades cannot see beyond the ends of their noses, possibly because of the greater political environment, possibly because of their arrogance.（刘宇昆译）

本例为针对红岸计划的政治批示，运用了多个具有中国特色的习惯用语和成语，如"胃口""扣帽子""一叶障目""自以为是"等。译者在翻译过程中没有简单地保留原文中的形象化表达，而是根据目标语言的使用习惯和文化背景进行了调整，提高了译文的流畅度和易读性，体现了译者的归化倾向。

练 习 题

一、基础练习：请翻译以下句子

1. 一天叶文洁值夜班。这是最孤寂的时刻。在静静的午夜，宇宙向它的聆听者展示着广漠的荒凉。

2. 叶文洁最不愿意看的，就是显示器上缓缓移动的那条曲线，那是红岸接收到的宇宙电波的波形。无意义的噪声。

3. 叶文洁感到这条无限长的曲线就是宇宙的抽象，一头连着无限的过去，另一头连着无限的未来，中间只有无规律无生命的随机起伏。

4. 一个个高低错落的波峰就像一粒粒大小不等的沙子，整条曲线就像是所有沙粒捧成行形成的一堆沙漠，荒凉寂寥，长得更令人无法忍受。你可以沿着它向前向后走无限远，但永远找不到归宿。

二、拓展练习：请翻译以下段落

1. 地球的变轨加速就这样年复一年地进行着。每当地球向远日点升去时，人们的心也随着地球与太阳距离的日益拉长而放松；而当它在新的一年向太阳跌去时，人们的心一天天紧缩起来。每次到达近日点，社会上就谣言四起，说太阳氦闪就要在这时发生了；直到地球再次升向远日点，人们的恐惧才随着天空中渐渐变小的太阳平息下来，但又在酝酿着下一次的恐惧……①

2. 人类的精神像在荡着一个宇宙秋千，更适当地说，在经历着一场宇宙俄罗斯轮盘赌：升上远日点和跌向太阳的过程是在转动弹仓，掠过近日点时则是扣动扳机！每扣一次时的神经比上一次更紧张，我就是在这种交替的恐惧中度过了自己的少年时代。其实仔细想想，即使在远日点，地球也未脱离太阳氦闪的威力圈，如果那时太

---

① 刘慈欣. 流浪地球[M]. 北京：中国科学技术出版社，2022.

阳爆发,地球不是被气化而是被慢慢液化,那种结果还真不如在近日点。①

第二节　航空航天文学翻译汉译英的方法

一、借　用

由于汉语和英语在书写系统上并不相同,加上文化交流的不平衡,汉译英中的借用主要体现在语音借用上,比如 kung fu(功夫)、dim sum(点心)等,而不会直接出现中文词汇或短语。这类借用词在英语中已经被广泛使用,并且逐渐成为英语的一部分。

在航空航天领域中,采用借词法进行汉英翻译的例子主要集中在中国政府部门以及官方媒体对于中国航空航天设施和专有名词的翻译上,比如神舟十三号就被译为 Shenzhou XIII,从而保证信息传递准确无误。此外,对于"中国宇航员"一词,一般会使用 taikonauts 一词,该词就是由中文"太空"的拼音"taikong"和英语 astronauts 的后缀组合而成的。

同时,对于人名译者大多也保留汉语拼音进行音译,但会适当增添注释,帮助读者理解,如下例:

例 3　中文原文:
"你好,我是墨子。"他自我介绍道。
"我是海人,你好。"
"啊,我知道你!"墨子兴奋地说,"在 137 号文明中,你追随过周文王。"②

英文译文:
"Hello," the man said. "I'm Mozi."
"Hello, I'm Hairen."
"Ah, I know you!" Mozi grew excited. "You were a follower of King Wen back in Civilization Number 137." (Translator's Note: Mozi was the founder of the Mohist school of philosophy during the Warring States Period. Mozi himself emphasized experience and logic, and was known as an accomplished engineer and geometer.)(刘宇昆译)

二、仿　造

相较于借用,仿造法在中英翻译中更为普遍。对于许多在目的语当中缺失的概

---

① 刘慈欣. 流浪地球[M]. 北京:中国科学技术出版社,2022.
② 刘慈欣. 三体[M]. 重庆:重庆出版社,2008.

念,译者都是将中文的各元素直译到英语中,构成新词。

在刘慈欣的《三体》中,作者自创了许多新概念,译者就运用了仿造法灵活翻译。比如,在遇到与"三"相关的概念时,译者采用了前缀 tri-,这一词缀表示"三,三个",常见的派生词有 triangle(三角形)、trilingual(三语者)。"三日凌空"描述的是空中有三个太阳,与太阳相关的词根为-solar,因此该词译为 trisolar。"三体人"指生活在三体世界的人,致力于发现三个太阳运动的规律,表示"某地居住的人"的词缀是-an,所以该词译为 trisolaran。同样的道理,"三体舰队"就译为 the Trisolaran Fleet,"三体文明"译为 the Trisolaran civilization,"三体时"译为 Trisolaran hour。

此外,对于"智子"这一概念,译者也进行了仿造。这一词汇中的"智"在汉语中用来表示"智慧",于是译者结合了古希腊语中表示智慧的词根 sophia。这一概念也与物理中的"质子"(proton)谐音,于是译者借鉴了表示"微观粒子"的后缀-on,最后将该词译为 sophon。

类似进行仿造的词汇还有"反动学术权威"(reactionary academic authorities)、"红旗战斗队"(Red Flag Combat Team)等,中英文基本对应。不过由于这些表达涉及文化信息,译者有时会采用加注等手段进行解释,既可以使读者了解指称对象的内容,同时也不会失去名称本身的异域色彩。

三、直 译

直译法在常规的对话翻译中比较常见,译文基本不对原语的结构进行改变。不过,由于中英文之间存在差异,在文学翻译中,大多数译文基本无法保留原文顺序。尽管如此,译者也会尽量做到遵循原文的意义和意象,在不影响理解的情况下采用直译。

例4 中文原文:
黑暗森林威慑是悬在两个世界头上的达摩克利斯之剑,罗辑就是悬剑的发丝,他被称为执剑人。①

英文译文:
Dark forest deterrence hung over two worlds like the Sword of Damocles, and Luo Ji was the single hair from a horse's tail that held up the sword. Thus, he came to be called the Swordholder.(刘宇昆译)

本例中,原文中达摩克利斯之剑的故事对于西方读者来说并不陌生,因此,对于"达摩克利斯之剑""执剑人"这些表述,可以按照原文进行直译,并不会给读者阅读带来困难。

例5 中文原文:
"汪教授,'科学边界'是个由国际顶尖学者构成的组织,对它的调查是件极其复

---

① 刘慈欣.三体[M].重庆:重庆出版社,2008.

杂和敏感的事情,我们真的是如履薄冰……"①

英文译文:

"Professor Wang, the Frontiers of Science is made up of elite international scholars. Investigating it is an extremely complex and sensitive matter. For us, it's like walking across thin ice."(刘宇昆译)

本例中,原文中"如履薄冰"表示好像走在很薄的冰面上一样,指"非常小心",这一比喻生动形象,不会给读者理解带来太大困难。所以译文中采用直译,保留了这一意象。

不过也有一些表达看似在英语和汉语中都存在相同的意象,但若是直接翻译反而会造成误译,例如:

例 6　中文原文:

汪淼点点头,比起宇宙闪烁来,他宁愿接受这个超自然。但沙瑞山立刻抽走了他怀中这唯一的一根救命稻草。②

英文译文:

Wang nodded. Compared to the idea of the universe flickering, he would prefer a supernatural saboteur. But Sha then deprived him of this last glimmer of hope.(刘宇昆译)

本例中,救命稻草指的是在即将绝望之时的一线生机、一丝希望。英语中有一句谚语叫作 A drowning man will catch at a straw,指溺水者不会放过身边任何能活命的机会,即使是一根稻草。它的意思与"病急乱投医"的意思相近,强调人在危机之中可能因慌乱而做出错误选择。所以放在这里并不合适,应当舍弃比喻的意象,译为 last glimmer of hope,即"一线生机"。

四、置　换

1. 动词的转换

例 7　中文原文:

老刀去幼儿园咨询的时候,着实被吓到了。③

英文译文:

Lao Dao's research on kindergarten tuition had shocked him.(刘宇昆译)

汉语强调动态,动词非常丰富,使用频率较高,而英语强调静态,较多使用名词。例 7 中,由于中英读者的语言使用存在差异,译者将动词转换为名词,增强了表述的

---

① 刘慈欣. 三体[M]. 重庆:重庆出版社, 2008.
② 同①。
③ 郝景芳. 北京折叠[M]. 南京:江苏凤凰文艺出版社, 2016.

客观性和静态感,符合英语读者的语言习惯。

2. 形容词的转换

例 8　中文原文:

马钢看上去<u>很兴奋</u>,让白记者注意到的人都这样,能在《大生产报》的通讯报道上露一下脸也是<u>很光荣</u>的事。①

英文译文:

Ma Gang's <u>excitement</u> was typical of most people Bai paid attention to. To be featured in the Great Production News would be a considerable <u>honor</u>. (刘宇昆译)

例 9　中文原文:

我又引入了第三个球体,情况发生了<u>令我震惊的</u>变化。②

英文译文:

I then introduced a third sphere, and <u>to my astonishment</u>, the situation changed completely. (刘宇昆译)

例 8 和例 9 中,译者将原文中的形容词在译文中都转化为名词。不过结合具体句子分析还可以发现,虽然转换后的词性相同,但变化后的内容在译文中充当的成分并不相同,它可以做主语、宾语、状语等多种成分。

五、调　适

1. 具体转为抽象

例 10　中文原文:

老刀要去第一空间送一样东西,<u>送到了</u>挣十万块,带来<u>回信</u>挣二十万。③

英文译文:

Lao Dao's errand required him to deliver a message to First Space—<u>success</u> would earn him a hundred thousand yuan, and if he managed to bring back a <u>reply</u>, two hundred thousand. (刘宇昆译)

本例中,原文"送到了"是具体动作,success 是抽象的结果;"回信"是具体指称,reply 是抽象概括。

---

① 刘慈欣. 三体[M]. 重庆:重庆出版社,2008.
② 同①。
③ 郝景芳. 北京折叠[M]. 南京:江苏凤凰文艺出版社,2016.

例 11　中文原文：

回到家时天已经黑了，我摸索着开了锁推门进去，开灯后<u>看到了</u>那熟悉的一切。①

英文译文：

It was already dark when I arrived, so I had to feel around to turn the lock and make my way in. Turning on the light <u>revealed</u> a familiar scene.（Joel Martinsen 译）

本例中，"看到"说的是具体动作，reveal 隐化了这一动作，同时这里说打开灯后看到家里的景象，reveal 一词的表达效果与由暗到亮的表述十分贴切。

2. 主动转为被动

例 12　中文原文：

这些娃们在烧香，接着他们又烧起纸来，火光<u>把</u>娃们的形象以橘红色在冬夜银灰色的背景上显现出来，这使他又想起了那灶边的画面。他脑海中还出现了另外一个类似的画面。（刘慈欣，《乡村教师》）

英文译文：

They were burning incense and paper, and their faces <u>were lit</u> red in the firelight against the silver-gray night. He <u>was reminded of</u> the sight of the children by the stove. Another scene emerged from the pool of his memory. (Adam Lanphier 译)

本例中，原文为火光把孩子的脸照亮了，译文中化主动为被动，更贴近英语读者的语言使用习惯。同时 their faces 作为主语，能够与前面的 they 实现更好的衔接，读起来不会有跳跃感。

3. 人称转为物称

例 13　中文原文：

把这幅画放进箱子前<u>我瞟了一眼画面，目光立刻被钉死在上面</u>。②

英文译文：

When I glanced at it before putting it into the box, <u>it seized my whole attention</u>.（Joel Martinsen 译）

例 14　中文原文：

老刀开始明白了。③

---

① 刘慈欣. 球状闪电[M]. 成都：四川科学技术出版社，2005.
② 同①.
③ 郝景芳. 北京折叠[M]. 南京：江苏凤凰文艺出版社，2016.

英文译文:

Understanding began to dawn on Lao Dao.(刘宇昆译)

例13中,关注点由"我的目光"变为"画",例14中,关注点由"老刀"变为"明白",从强调人到强调物,属于人称转物称,符合目的语读者的语言使用习惯。

4. 调整对象顺序

例15　中文原文:

我必须要面对它了,因为开学后,大气电学专业的课程就要开始了。讲大气电学的是一名叫张彬的副教授,这人五十岁左右,个子不高不矮,眼镜不薄不厚,讲话声音不高不低,课讲得不好不坏。①

英文译文:

Classes in atmospheric electricity started that semester, meaning I would finally have to face it. The subject was taught by an assistant professor named Zhang Bin. He was about fifty, neither short nor tall, wore glasses that were neither thick nor thin, had a voice that was neither loud nor soft, and his lectures were neither great nor terrible.(Joel Martinsen 译)

本例中,原文描述的对象按照先后顺序分别为"我""课程""副教授",接着又对副教授的年龄、身高、眼镜、声音、讲课进行描述,采用流水句,体现了汉语母语者的语言和思维习惯。译者对文本信息进行整合,对行文顺序进行重新调整,旨在使译文符合英语使用习惯,使文章更加通顺易懂。

5. 反向表述

例16　中文原文:

他不知道有什么理由拒绝。②

英文译文:

He could think of no reason to not take up the offer.(刘宇昆译)

例17　中文原文:

只是,这座水塔是在我考上大学之后才建成的,我两年前离开时,塔身只在脚手架中建了一半。③

英文译文:

Except, the water tower had not been completed until after I went off to college. When I left two years ago, it had been half-finished and covered

---

① 刘慈欣. 球状闪电[M]. 成都:四川科学技术出版社, 2005.
② 同①。
③ 同①。

in scaffolding. (Joel Martinsen 译)

例 18　中文原文：

汪淼第一眼就对来找他的警察没有好感。①

英文译文：

As soon as Wang saw the cops, he felt annoyed. (刘宇昆译)

例 16、例 17 和例 18 中，译者采用了相反的表述方式传达了相同的语义，若不进行反向表述，译文就会读起来难以理解，不符合英语的表达习惯。

6. 用词变化

例 19　中文原文：

对面动物园大门旁的一排霓虹灯中有一根灯管坏了，不规则地闪烁着；近处的一棵小树上的树叶在夜风中摇动，反射着街灯的光，不规则地闪烁着；远处北京展览馆俄式尖顶上的五角星也在反射着下面不同街道上车灯的光，不规则地闪烁着……②

英文译文：

By the entrance of the zoo across the street, there was a row of neon lights. One of the lights was about to burn out and flickered irregularly. Nearby, a small tree's leaves trembled in the night breeze, twinkling without pattern as they reflected streetlight. In the distance, the red star atop the Beijing Exhibition Center's Russian-style spire reflected the light from the cars passing below, also twinkling randomly…（刘宇昆译）

原文中出现了三次"不规则地闪烁着"，而译文采用的表达各不相同，分别译为 flickered irregularly, twinkling without pattern, twinkling randomly。汉语喜欢重复，而英语不喜重复，倾向于采用不同词汇进行替换。在这里若保留原词，读起来就会感觉啰嗦累赘，失去美感。

例 20　中文原文：

"它是三体纪念碑，也是一个墓碑。"秘书长仰望着半空中的摆锤说，从这里看去，它足有一个潜水艇那么大。

"墓碑？谁的？"

"一个努力的，一个延续了近二百个文明的努力，为解决三体问题的努力寻找太阳运行规律的努力。"③

---

① 刘慈欣. 三体[M]. 重庆：重庆出版社，2008.

② 同①。

③ 同①。

英文译文：

"It's a monument for Trisolaris, as well as a tombstone." The secretary general looked up at the pendulum. From down here, it appeared as big as a submarine.

"A tombstone? For who?"

"For an aspiration, a striving that lasted through almost two hundred civilizations: the effort to solve the three-body problem, to find the pattern in the suns' movements."（刘宇昆译）

原文中出现了四次"努力"，而译文中将其转化成 aspiration，striving，effort，灵活变换词汇，更加符合英语读者的阅读习惯。

六、对　等

例 21　中文原文：

血液复苏的小腿开始刺痒疼痛，如百爪挠心，几次让他摔倒，疼得无法忍受，只好用牙齿咬住拳头。①

英文译文：

As circulation returned to his numb leg, his calf itched and ached as though he was being bitten by thousands of ants. Several times, he almost fell. The pain was intolerable, and he had to bite his fist to stop from screaming.（刘宇昆译）

本例中，汉语"百抓挠心"用来形容主人公小腿受伤后刺痒疼痛的感觉，若直译此处意象，会增添"爪""心"等内容，增加读者理解的难度，因此采用 as though he was being bitten by thousands of ants，虽形式不同，但同样传神。

例 22　中文原文：

他爸爸是个顽固的飞船派，因参加一次反联合政府的暴动，现在还被关在监狱里。有其父必有其子。②

英文译文：

His father was an incorrigible member of the Spaceship Faction, and he was, in fact, still in prison for joining an insurgency against the Unity Government. Seeing Tung, I guessed that the apple hadn't fallen far from the tree.

本例中，"有其父必有其子"的意思是父母什么样，孩子就什么样，译文中将其转化为 the apple hadn't fallen far from the tree，用英语读者更熟悉的描述表示相同的语义，读起来更加自然易懂。

---

① 郝景芳.北京折叠[M].南京：江苏凤凰文艺出版社，2016.
② 刘慈欣.流浪地球[M].北京：中国科学技术出版社，2022.

例 23　中文原文：

一些同志现在是一叶障目，有大环境的原因，也有很多人是自以为是。①

英文译文：

Some comrades cannot see beyond the ends of their noses, possibly because of the greater political environment, possibly because of their arrogance.（刘宇昆译）

本例中，成语"一叶障目"的意思是一片叶子挡在眼前就让人看不见外面广阔的世界，比喻被暂时的现象所蒙蔽。译者将其转换为 cannot see beyond the ends of one's nose，同样用来形容人目光短浅。

七、显　化

1. 明晰指代

例 24　中文原文：

在城市边缘的那所著名大学的操场上，一场几千人参加的批斗会已经进行了近两个小时。在这个派别林立的年代，任何一处都有错综复杂的对立派别在格斗。②

英文译文：

At the edge of the city, on the exercise grounds of Tsinghua University, a mass "struggle session" attended by thousands had been going on for nearly two hours.（刘宇昆译）

本例中，原文为"那所著名大学"，作者特意没有明示出是哪所学校，而是给出"位于城市边缘"这一描述，给读者留下探索空间。后文中提到，叶文洁"回到这座城市，在父亲曾工作过的大学中讲授天体物理学直到退休"，而且她是"清华大学物理系天体物理专业教授"，由此可以推断出本句中执行批斗的地点是清华大学的操场。然而英语母语者的相关背景知识比较有限，如果保留本句中的代词，读者可能并不会猜到是清华大学，无法在人物之间建立清楚的逻辑关系，所以译者显化了指代的内容，降低了读者阅读和理解的难度。

例 25　中文原文：

我和它，像两个要用一生时间准备一场决斗的骑士，当我没准备好的时候，既不去见它也不去想它。③

英文译文：

My fascination and I were two knights whose entire lives would be

---

① 刘慈欣. 三体[M]. 重庆：重庆出版社，2008.
② 同①.
③ 刘慈欣. 球状闪电[M]. 成都：四川科学技术出版社，2005.

devoted to preparing for a single duel, and until I was ready I would neither think about it nor seek it out directly. (Joel Martinsen 译)

本例中,原文采用代词,译者将"它"的具体指称明示出来,译为 fascination,使表述更为清晰。

2. 明确逻辑衔接词

例 26　中文原文:

红岸发射系统的发射器是超高功率的设备,全部使用"文革"期间生产的国产元件,由于质量不过关,故障率很高,不得不在每十五次发射后就全面检修一次,每次检修完成后都要例行试运行,参加这种发射的人很少,目标和其他发射参数也是比较随意的。①

英文译文:

The Red Coast transmitter was ultra-high-powered, but all of its components were domestically produced during the Cultural Revolution. As the quality of the components was not up to par, the fault rate was very high. After every fifteenth transmission, the entire system had to be overhauled, and after each overhaul, there would be a test transmission. Few people attended these tests, and the targets and other parameters were arbitrarily selected. (刘宇昆译)

本例中,原文的逻辑关系并不凸显,但在英语译文中,译者增加了 but, as, and 等表示分句间逻辑关系的词,显化了表达的逻辑。

3. 借助标点

例 27　中文原文:

叶文洁走到为她空出的一圈空地中央,举起一只瘦削的拳头,令汪淼不敢想象的事情出自她的力量和坚定地说:"消灭人类暴政!"②

英文译文:

Ye stood in the middle of the space the crowd cleared for her, raised a bony fist, and—with a resolve and strength that Wang could not believe she possessed—said, "Eliminate human tyranny!" (刘宇昆译)

本例中,译者增加了原文中没有的标点符号,为信息划分出了层次,明确区分出了主要信息和次要的修饰成分。

---

① 刘慈欣. 三体[M]. 重庆:重庆出版社, 2008.
② 同①。

4. 明晰语义

例28　中文原文：

学者们发现，与大多数人美好的愿望相反，人类不可能作为一个整体与外星文明接触，这种接触对人类文化产生的效应不是融合而是割裂，对人类不同文明间的冲突不是消解而是加剧。①

英文译文：

Scholars found that, contrary to the happy wishes of most people, <u>it was not a good idea</u> for the human race as a whole to make contact with extraterrestrials. The impact of such contact on human society would be divisive rather than uniting, and would exacerbate rather than mitigate the conflicts between different cultures.（刘宇昆译）

本例中，原文为"不可能"，但是译者将这里翻译成 it was not a good idea。从逻辑上看，这句并非描述人类与外星文明接触的难度，而是强调两种文明接触后会产生消极影响。译文将隐含的语义明示出来，避免产生误解。

5. 弥补知识空缺

例29　中文原文：

"不到十分钟吧，告诉你，我是连里最快的油锯手，我到哪个班，<u>流动红旗</u>就跟我到那儿。"②

英文译文：

"No more than ten minutes. Let me tell you, I'm the fastest chain saw operator in the company. Whichever squad I'm with, <u>the red flag for model workers follows me</u>."（刘宇昆译）

对于中文读者来说，对于"流动红旗"并不陌生，但对于英语读者来说，red flag 只是一个具体事物，因此可能无法理解这句话提到红旗的用意何在，不知所云。因此译文解释出 model worker，弥补了知识空缺，从而减小了读者理解的难度。

例30　中文原文：

但这时，绍琳却做出了一件出人意料的事，与一位受迫害的教育部高干结了婚，当时那名高干还在干校住"牛棚"劳改中。③

英文译文：

But then Shao did something that no one expected. She married a

---

① 刘慈欣. 三体[M]. 重庆：重庆出版社，2008.
② 同①.
③ 同①.

persecuted high-level cadre from the Education Ministry. At that time, the cadre still lived in a "cowshed" for reform through labor. (Translator's Note: "Cowsheds" were locations set up by work units (factories, schools, towns, etc.) during the early phases of the Cultural Revolution to detain the counter-revolutionary "Monsters and Demons" (reactionary academic authorities, rightists, the Five Black Categories, etc.) at the work unit.)(刘宇昆译)

本例中,译者对"牛棚"加注进行解释。不了解中国社会历史文化背景的外国读者可能难以明白为什么要住在牛棚里,从而导致理解困难,译者增加注释,弥补了相关的背景知识,利于读者理解。

例31 中文原文:

与其他牛鬼蛇神相比,反动学术权威有他们的特点:当打击最初到来时,他们的表现往往是高傲而顽固的,这也是他们伤亡率最高的阶段;他们有的因不认罪而被活活打死,有的则选择了用自杀的方式来维护自己的尊严。①

英文译文:

Compared to other "Monsters and Demons," reactionary academic authorities were special: During the earliest struggle sessions, they had been both arrogant and stubborn. That was also the stage in which they had died in the largest numbers. Over a period of forty days, in Beijing alone, more than seventeen hundred victims of struggle sessions were beaten to death. Many others picked an easier path to avoid the madness: Lao She, Wu Han, Jian Bozan, Fu Lei, Zhao Jiuzhang, Yi Qun, Wen Jie, Hai Mo, and other once-respected intellectuals had all chosen to end their lives (Translator's Note: These were some of the most famous intellectuals who committed suicide during the Cultural Revolution. Lao She: writer; Wu Han: historian; Jian Bozan: historian; Fu Lei: translator and critic; Zhao Jiuzhang: meteorologist and geophysicist; Yi Qun: writer; Wen Jie: poet; Hai Mo: screenwriter and novelist.)(刘宇昆译)

《三体》故事的背景是"文化大革命",而英语读者对此了解可能很少,不清楚这场运动涉及的范围有多大,究竟有什么影响,因此可能也就无法理解后面的故事情节。在这里,译者加注进行了相当长的解释,显化甚至增加了内容,弥补了读者在这方面的知识空缺。

---

① 刘慈欣. 三体[M]. 重庆:重庆出版社,2008.

八、隐 化

例 32 中文原文：

十五岁少女的胸膛是那么柔嫩，那颗子弹穿过后基本上没有减速，在她身后的空中发出一声啾鸣。①

英文译文：

Her fifteen-year-old body was so soft that the bullet hardly slowed down as it passed through it and whistled in the air behind her.（刘宇昆译）

例 33 中文原文：

她父亲留下了一堆唱片，她听来听去，最后选择了一张巴赫的反复听，那是最不可能令孩子，特别是女孩子入迷的音乐了。②

英文译文：

Her father left behind some records. She listened to all of them and finally picked something by Bach as her favorite, listening to it over and over. That was the kind of music that shouldn't have mesmerized a kid.（刘宇昆译）

《三体》中有一些内容在目的语读者看来可能带有对女性的刻板印象，译者对这些内容进行了淡化或删除，如例 32 将胸膛转化为 body，例 33 将"特别是女孩子"这一内容删除。

九、改 编

例 34 中文原文：

但十五一过，村里的青壮年都外出打工挣生活去了，村子一下没了生气。（刘慈欣，《乡村教师》）

英文译文：

But as soon as the Spring Festival ended, all the youths of the village left again to look for work, and the place fell back into torpor.（Adam Lanphier 译）

本例中，前文提到过年时平常在外打工的人会回到村子，变得热闹起来。本处描述他们过完年离开，村子没了生气。英语母语者可能了解中国人在春节时会回家庆贺，但对于正月十五以后多数人会返工的情况却并不了解。若按照原文意义进行翻译，则需要解释中国农历新年和返工相关的诸多内容，不利于阅读的流畅性。为了方便读者理解，译者将此处理为春节过后村子又变得冷清。

---

① 刘慈欣. 三体[M]. 重庆：重庆出版社，2008.
② 同①。

除了对于个别句段的改编,译者可能还会对作品整体进行较大的调整。比如,《三体》中文版采用了分叙的手法,存在几条相互交叉的时间线,并且存在插叙等。而译者提出,由于注意到中外文学在叙事技巧上不同,读者的期望和偏好存在差异,因此在保证译文意义与原文意义一致的基础上调整了部分叙事技巧。经过重新编排的作品被分为三个章节,分别为"寂静的春天"、"三体"和"人类的黄昏"。翻译版本的第一章便是"疯狂年代",在开篇便介绍宏大的社会历史场面,制造紧张氛围,充满戏剧性,紧紧抓住读者的注意力。

练 习 题

一、基础练习:请翻译以下句子

1. 这场漫长的战争伴随着整个人类文明,现在仍然胜负未定,虫子并没有被灭绝,它们照样傲行于天地之间,它们的数量也并不比人类出现前少。

2. 一到社会上,才发现自己是个地地道道的废物,除了数学啥也不会,在复杂的人际关系中处于半睡眠状态,越混越次。

3. 我每天夜里都在一两点才回到宿舍,听着某个室友在梦中喃喃地念着女朋友的名字,这才意识到还有另一种生活。

二、拓展练习:请翻译以下段落

1. 太阳的灾变将炸毁和吞没太阳系所有适合居住的类地行星,并使所有类木行星完全改变形态和轨道。自第一次氦闪后,随着重元素在太阳中心的反复聚集,太阳氦闪将在一段时间反复发生,这"一段时间"是相对于恒星演化来说的,其长度可能相当于上千个人类历史。①

2. 有一天,新闻报道海在融化,于是我们全家又到海边去。这是地球通过火星轨道的时候,按照这时太阳的光照量,地球的气温应该仍然是很低的,但由于地球发动机的影响,地面的气温正适宜。能不穿加热服或冷却服去地面,那感觉真令人愉快。地球发动机所在的这个半球天空还是那个样子,但到达另一个半球时,真正感到了太阳的临近:天空是明朗的纯蓝色,太阳在空中已同启航前一样明亮了。②

---

① 刘慈欣. 流浪地球[M]. 北京:中国科学技术出版社,2022.
② 同①。

第五章 航空航天文学翻译批评与赏析

航空航天文学是兼具科技与文学特点的一类文本,因此在赏析翻译作品时应注意全面性,从多种角度来进行全方位评析。本章内容为航空航天文学翻译批评与赏析,共有三节:第一节将简单综述翻译批评的研究历史情况以及在翻译批评中常用的一些翻译质量评估模型;第二节和第三节则分别从航空航天类文学"英译汉"和"汉译英"两个角度对翻译作品进行具体批评实践,以期能为翻译活动提供一些建议。

第一节 文学翻译批评与赏析

本节主要包含两部分内容:翻译批评研究综述和翻译质量评估模型综述。在完成本节阅读后,读者能对翻译批评的研究历史有基本了解,同时也能对现有的一些常用评估模型有所认识。本节所提到的以文本类型为倾向的赖斯翻译质量评估模型也被运用在了后两节的具体实践中。

一、翻译批评研究综述

20世纪60年代以来,西方的翻译研究蓬勃发展。以解构主义、女性主义、多元系统理论、描写翻译学、操纵学派等为代表的文化学派促使在20世纪90年代期间,翻译研究出现了向文化转向的特征。这些翻译理论摆脱了传统翻译研究的特征,理论家们把翻译视为一项与文化密切结合的实践活动,提出翻译研究也应关注译者、译作等语言以外的可能会影响翻译的因素。因此,在进行翻译批评时,不仅要关注语言之内的因素,如语言组织形式、词汇使用方式、句型构成等,还需要将文本放在意识形态、社会文化体系等大环境中进行描写研究。

1972年,詹姆斯·霍姆斯发表了一篇名为《论翻译研究的名与实》的文章,本篇文章第一次较为详实地定义了翻译学的范畴,同时也规范了翻译学研究的领域。自此,翻译研究逐渐成为学术界的热点问题。根据霍姆斯对翻译研究的分类,翻译研究包括描写、理论、应用三大模块,翻译批评则是应用模块中的重要组成部分。翻译批评在翻译学研究中占据了不可忽视的地位,但对翻译批评的理论研究却相对"缓慢"。纽马克指出:"翻译批评是翻译理论与实践之间的一条根本纽带""是翻译研究的重要组成部分"[1]。翻译学在不断发展壮大的过程中,不但要关注翻译批评的实践,还需

---

[1] Newmark Peter. A Textbook of Translation[M]. London: Prentice Hall, 1988: 184.

要进一步完善健全翻译批评的理论体系。

近三十年来,国内外翻译界开始关注翻译批评的理论问题,对翻译批评的任务、目的、方法提出了一些重要观点。《中国翻译词典》中"翻译批评"的条目中指出:"从广义上讲,翻译批评即参照一定的标准,对翻译过程及其译作质量与价值进行全面的评价,评价的标准因社会历史背景而异,评价的目的在于促使译作最大限度地忠实于原作,并具有良好的社会价值。现代翻译批评理论认为,全面的翻译批评应当包括五个方面的内容:分析原作,着重了解作者的意图与原作具有的功能;分析译者翻译原作的目的、所采用的翻译方法及其译作针对或可能吸引的读者对象;从原作与译作中选择有代表性的文字进行详细的对比研究;从宏观与微观的角度评价译作,包括译者采用的技巧与译作的质量等方面的内容;评价译作在译语文化或学科中的作用与地位。"①《译学词典》中"翻译批评"词条的内容是:"翻译批评是一种具有一定实践手段和理论目标的精神活动,是从一定的价值观念出发,对具体的翻译现象(包括译作和译论)进行分析和评价的学术活动,是审美评价与科学判断的有机统一。翻译批评的任务是以一定的翻译标准为准绳,以科学的方法对译本或译论的艺术价值或科学价值进行判断,对其不足之处进行理论上的鉴别,特别要检视翻译实践的跨文化交际效果,从中探索译者的审美境界、科学视野和艺术技巧,以提高译者和读者的鉴别能力。翻译批评对促进文化事业和翻译事业的健康发展起到积极的作用。"②许钧在《翻译论》中写道:"从广义上讲,翻译批评就是'理解翻译与评价翻译'。从狭义上讲,翻译批评是对翻译活动的理性反思与评价,既包括对翻译现象、翻译文本的具体评价,也包括对翻译本质、过程、技巧、手段、作用、影响的总体评析。就翻译批评的目的而言,它并不仅仅在于对具体译作或译法作裁判性的是非判别,更在于对翻译活动何以进行、如何进行加以反思与检讨,进而开拓翻译的可行性,促进翻译活动健康而积极地发展,体现翻译活动所具备的各种价值,真正起到翻译活动应有的作用。"③以上学者给翻译批评所下的定义都或多或少地注意到了翻译的规定性和描写性特征,同时又都提到了翻译的价值问题。在前人的理论基础上,翻译批评研究蓬勃发展。

我国的翻译批评史可以分成三个阶段:第一个阶段是20世纪以前的翻译评论;第二个阶段是20世纪初的翻译批评;第三个阶段是1949年以来的当代翻译批评。④三国时期的支谦在《法句经序》中记录的一次关于翻译原则的讨论可视为我国最早的翻译评论,此为第一阶段的代表作。清末民初,随着翻译事业的繁荣,翻译评论应运而生。王国维在1906年写的《书辜氏汤生英译〈中庸〉后》中对辜鸿铭的译文提出了批评,并阐明了自己对古书外译的观点。后来,鲁迅、郭沫若、郑振铎、茅盾、郁达夫等

---

① 林煌天. 中国翻译词典[M]. 武汉:湖北教育出版社,1997:184.
② 方梦之. 译学词典[M]. 上海:上海外语教育出版社,2004:346.
③ 许钧. 翻译论[M]. 武汉:湖北教育出版社,2003:403.
④ 王恩冕. 论我国的翻译批评:回顾与展望[J]. 中国翻译,1999(4):7-10.

学者也纷纷对翻译实践进行评析。此为第二阶段的整体现象。进入第三阶段后,翻译批评的发展也经历了几个颇具历史意义的事件:1954年茅盾先生在全国文学翻译工作会议报告中专门论述了翻译批评,并且对翻译批评作出了最高要求。20世纪50年代中期对翻译标准大讨论推进了翻译批评的发展。1995年《读书》《文汇读书周刊》《光明日报》等报刊就《红与黑》的不同译本展开了广泛而热烈的讨论。《中华读书报》也在1997年发表了"翻译作品面面观"的征文活动等,为以后翻译批评的深入发展奠定了较好的基础。

翻译批评研究发展至今,许多专门论述翻译批评的代表著作也陆续出版。国内专门论述翻译批评的代表著作有十多部。许钧先生在《文学翻译批评研究》中归纳出了六种翻译批评的方法:逻辑验证的方法、定量定性分析方法、语义分析的方法、抽样分析的方法、不同翻译版本的比较和佳译赏析的方法[1];喻云根先生在《英美名著翻译比较》中对12部英美名著或其片段的不同汉译本进行了比较研究,对各个译本分别从思想内容到语言风格进行了较全面的剖析,可谓英译汉名著多译本比较的权威著作[2];姜治文、文军先生的《翻译批评论》[3]是一本论文集性质的翻译批评著作,书中不但有翻译批评的理论阐释,还有对译品、译者的评价;周仪、罗平先生的《翻译与批评》[4]从翻译标准、翻译批评及译文比较与赏析这三个角度详细论述了翻译批评中所涉及的诸多问题;马红军先生的《翻译批评散论》[5]对名译进行对比后提出自己的异议;奚永吉先生的《文学翻译比较美学》[6]从比较美学角度借鉴我国古今文论、诗论、曲论、画论中的美学原理对译本加以详细比较分析,充分挖掘译本的美学价值;方梦之先生在《翻译新论与实践》[7]中阐述了译学研究的四种具体方法:对比、调查、描写、论证;冯庆华先生的《文体翻译论》[8]从文体的角度研究翻译风格,是多种译本比较和对比研究的典范之作。除了这些极具代表性的著作之外,还有许多国内学者为翻译批评研究提供有益的理论基础和实践基础。郑海凌先生的《文学翻译学》[9]提出"和谐"说的翻译批评标准;许钧、袁筱一先生的《当代法国翻译理论》[10]提出建构性的翻译批评取向。

西方的翻译批评研究,无论是在广度上还是在深度上都先国内一步。其中的一

---

[1] 许钧. 文学翻译批评研究[M]. 南京:译林出版社,1992:51-52.
[2] 喻云根. 英美名著翻译比较[M]. 武汉:湖北教育出版社,1996.
[3] 姜治文,文军. 翻译批评论[M]. 重庆:重庆大学出版社,1999.
[4] 周仪,罗平. 翻译与批评[M]. 武汉:湖北教育出版社,1999:14.
[5] 马红军. 翻译批评散论[M]. 北京:中国对外翻译出版公司,2000.
[6] 奚永吉. 文学翻译比较美学[M]. 武汉:湖北教育出版社,2001.
[7] 方梦之. 翻译新论与实践[M]. 青岛:青岛出版社,2001.
[8] 冯庆华. 文体翻译论[M]. 上海:上海外语教育出版社,2002.
[9] 郑海凌. 文学翻译学[M]. 郑州:文心出版社,2000:377-391.
[10] 许钧,袁筱一. 当代法国翻译理论[M]. 武汉:湖北教育出版社,2001:273-297.

个重要标志就是,关于翻译批评理论探讨的英语文献数量远远超出汉语同类作品①,更遑论其他语种了。鉴于西方翻译批评研究文献十分丰富,牵涉语种过多,在这里仅对西方极具代表性的五位学者的翻译批评研究进行简要的总结归纳和评论。

彼得·纽马克(Peter Newmark)是英国著名的翻译理论家,早在1988年出版的《翻译教程》一书中他就专用一个章节来讨论翻译批评问题。关于如何进行翻译批评,纽马克认为应该遵循以下五个步骤:(1)分析原文。对原作的分析应包括原作作者意图、原作文本类型、译作的读者群等因素。(2)分析译作。重点分析译者对原作作者意图的理解和译者所采用的翻译策略以及采用该翻译策略的原因。(3)比较分析原作和译作。这是翻译批评的核心内容。(4)评论译作质量。可从不同角度进行,最重要的是检查原作中重要的"恒定内容"(如主要观点、事实等)是否准确再现。(5)评论译作在译语文化中的价值。②

凯瑟琳娜·赖斯(Katharina Reiss)系德国翻译理论家,其经典力作《翻译批评:潜力与制约》已于1971年出版。她根据语言功能理论,建立了一个文本类型模式,从语篇类型和语篇功能的宏观层面提出了翻译的原则和批评标准等问题。赖斯将文本划分为四大类:(1)注重内容的文本(content-focused text);(2)注重形式的文本(form-focused text);(3)注重感染的文本(appeal-focused text);(4)以听觉等为媒介的文本(audio-medial text)。③ 赖斯认为,不同类型文本的译文应有各自相异的翻译和批评标准。

沃尔夫拉姆·威尔斯(Wolfram Wilss)也是一位德国翻译理论家,1977年他写作了《翻译学——问题与方法》一书。在书中第11章,他专门探讨了诸如翻译批评的处境及存在的问题和翻译批评客观化的可能性等问题,并提出了一个使翻译批评客观化的参照框架。④

朱莉安·豪斯(Juliane House)是国际翻译批评界第一个建构了系统、全面的翻译质量评估模式的学者。⑤ 豪斯的翻译质量评估思想主要体现在一脉相承的三本专著中。她认为,翻译质量评估应以翻译理论为前提,基于此,前者应是任何翻译理论的中心问题。对等既是翻译理论中重要的核心概念,也是翻译质量评估的基础。豪斯在此基础上提出了著名的豪斯翻译质量评估模型,该模型至今还在翻译批评研究中广泛使用。

---

① 胡德香. 翻译批评新思路:中西比较语境下的文化翻译批评[M]. 武汉:武汉出版社,2006:128.
② Newmark Peter. A Textbook of Translation[M]. Shanghai:Shanghai Foreign Language Education Press,2001:186-189.
③ Reiss Katharina. Translation Criticism:The Potentials & Limitations[M]. Trans. by Erroll F Rhodes. Shanghai:Shanghai Foreign Language Education Press,2004:27-47.
④ Wilss Wolfram. The Science of Translation:Problems and Methods[M]. Shanghai:Shanghai Foreign Language Education Press,2001.
⑤ 司显柱. 朱莉安·豪斯的"翻译质量评估模式"批评[J]. 外语教学,2005(3):79-84.

安托万·贝尔曼(Antoine Berman)是法国当代著名翻译理论家,其翻译批评研究专著《翻译批评论:约翰·唐》以海德格尔的阐释学和本雅明的批评理论为基础,对翻译批评的本质、目的、途径、标准等问题作了系统思考,具有"开创性意义"。① 贝尔曼指出,翻译批评具有积极意义。在他看来,"translation criticism"一词可能暗示对翻译内容的负面评价,但是"一个纯粹负面的批评不是真正的批评",因为翻译批评的终极目的是"评价译作的真实性"。②

基于以上国内外翻译批评研究的内容综述,我们可以发现,翻译批评的理论发展和实践技巧是多角度且多层面的。我们不仅可以从语言角度对比原文和译文的同异之处,探究译者极具想法的翻译技巧,还可以从非语言角度观察原文译作所处的背景环境,从而探索翻译实践的因与果。但需要注意的是,翻译批评并不是一项只进行"批评"的活动,正如贝尔曼所述,翻译批评活动的真谛在于评价译作的真实性。世上不会有绝对完美的译作,当我们研究一篇译作的优缺点时,不同的角度与观点就会带给我们不同的评价结果。尽量客观全方位地考虑原作与译作的关系与环境,真实地从多个方面探讨翻译实践的价值,这是一个评价者应做到的。

二、翻译质量评估模型综述

翻译批评的标准一直是学者们不断讨论的一个问题,在这样的背景下,不断有学者提出翻译质量评估模型,以期能够为翻译批评提供一个评估框架。

在翻译教育领域,翻译质量评估模型可以在学生进行翻译实践时,不断向学生提供反馈信息,并在教学结束时提供学生整体表现的总结性信息。③ 在翻译行业,翻译质量评估代表了质量掌控机制的重要组成部分,以确保能够以高标准的要求提供翻译服务。④ 在翻译研究中,翻译质量评估模型提供的指标经常用来验证或反驳研究的初始假设。⑤ 因此,我们不难发现,在对许多著名的翻译质量评估理论模型进行实践后,其结果在如何评估翻译的原则问题上也提供了或相反或互补的观点。尽管如此,翻译质量评估模型却很少在翻译测试和评估中大规模地使用。这可能是因为翻译研究学者倾向于从"理论和案例研究的角度,而不是大规模实验的角度"来对待翻

---

① 胡翠娥. 绘事后素:翻译批评理论与实践的有机结合:评安托万·贝尔曼《翻译批评探索:约翰·多恩》[J]. 中国翻译, 2015(1):63-67.

② Berman Antoine. Toward a Translation Criticism: John Donne[M]. Trans. by Francoise Massardier-Kenney. Kent: Kent State University Press, 2009:25.

③ Arango-Keeth F, Geoffrey S K. Assessing Assessment: Translator Training Evaluation and the Needs of Industry Quality Assessment[M]//Baer B J, Koby G S. Beyond the Ivory Tower: Rethinking Translation Pedagogy, Amsterdam: John Benjamins, 2003:117-134.

④ Martinez Mateo R. A Deeper Look into Metrics for Translation Quality Assessment (TQA): A Case Study[J]. Miscelanea, 2014(49):73-93.

⑤ Rothe-Neves R. Translation Quality Assessment for Research Purposes: An Empirical Approach[J]. Cuadernos De Traducao, 2008(2):113-131.

译质量评估。① 所以,翻译质量评估模型似乎更适合于翻译批评,而并不适用于需要评估相对大量译文的实际环境中。②

国外的翻译质量评估模型经过了几年的演变,也出现了各种不同的评估方法。在不同的环境与条件下,大家也会选择使用不同的模型以达到最佳效果。在本节中,我们主要介绍七种不同的评估模型,包括直观评估、错误分析、基于语料库的评估、以一定评分标准进行评分、使用混合方法进行评分、基于项目的评估、比较判断。

毋庸置疑,翻译质量评估中最古老的方法就是基于对翻译整体质量的直觉、经验和轶事判断,本质上是豪斯所说的翻译质量评估中数百年的"心灵论观点"。③ 一般情况下,评估是基于评估者的印象、感觉和个人喜好,而不是诉诸于任何一种有明确定义的标准。正如豪斯所观察到的那样,这种评估在逐渐发展成为一种新型解释性评估,即翻译文本的质量取决于"旁观者的眼睛"或评估者的个人解释。因此,由于缺乏客观性和一致性,以及无法生成有用的信息度量,这种方法的实用性受到了严重的限制。

对直觉评估高度主观性的不满,促使翻译研究人员寻求一种新的方法来进行翻译质量评估。在过去的几十年里,吸引了最多的学术关注,且在教育和专业环境中被广泛使用的方法是错误分析(error analysis)。这种方法以识别错误的概念为基石,规定了翻译中可以识别的不同类型的错误,然后对译者所犯错误的严重程度进行评级,并相应地赋予不同的权重④。这种有原则、有计划的翻译质量评估方法有许多好处。其一,由于有明确的质量标准、详细的错误类型和评估结果的限定,这种评估模式是客观、系统和可靠的;其二,在于其整个过程是透明的,因为评估程序是预先确定的,所有相关的利益相关者都会对其有所了解;其三,它能够对翻译的优势和劣势进行详细和细致的诊断⑤。在该类评估模型中,LISAQA 模型和 SAE J2450 模型是最具影响力的。

后来随着研究的不断推进,基于语料库的翻译质量评估也出现了。其最早于 21 世纪初由鲍克(Bowker)提出,他认为运用语料库可以在翻译质量评估中发挥更大作用。这类评估模式的原理是:首先定制一个相关的参考语料库,然后通过语料库产生一致性列表,并以这个列表来作为评判翻译质量的基准。评估者可以根据这些

---

① Doherty S. Issues in Human and Automatic Translation Quality Assessment[M]//Kenny D. Human Issues in Translation Technology. London, UK: Routledge, 2017: 131-148.

② McAlester G. The Evaluation of Translation into a Foreign Language[M]//Schaffner C, Adab B. Developing Translation Competence, Amsterdam: John Benjamins, 2000: 229-241.

③ House J. Translation Quality Assessment: Past and Present [M]//House J Translation: A Multidisciplinary Approach. London: Palgrave Macmillan, 2014: 241-264.

④ Williams M. The Assessment of Professional Translation Quality: Creating Credibility Out of Chaos [J]. Traduction, Terminologie, Redaction 1989 (2): 13-33.

⑤ Eyckmans J, Anckaert P, Segers W. Translation and Interpretation Skills[M]//Tsagari D, Banerjee J. Handbook of Second Language Assessment, Berlin: De Gruyter/Mouton, 2016: 219-235.

基准来比较和判断多个翻译版本中的单词和短语选择是否合适。① 该方法的主要目的是帮助评估者根据真实的、有经验的、可验证的语料库数据做出准确的判断。另一个推动基于语料库的评估模型发展的是克雷斯波(Jimenez-Crespo)。他对传统纠错型评估模式的理论严谨性和经验合理性提出了质疑,他认为这些错误类型学大多是基于经验和直觉。因此,克雷斯波主张采用以语料库为基础、以经验为依据的方法来微调错误分析方式,使其达到更高的一致性。② 虽然使用语料库是以证据为基础,其能代替错误评估模型中的一些不合理之处。但是广泛使用这个模型也会产生几个问题。首先,这个模型需要有技术专长的人来编纂参考语料库,并对使用者进行相关培训以适当使用工具;其次,语料库也需要时间、精力和专业知识来维护和扩展;最后,这个评估模型还可能需要统计知识来运行分析和正确解释结果。因此,基于语料库来进行评估的支持者建议,不能以这种方法取代传统的评估方法。相反,这是一个很有用的工具。做翻译批评时可以有效地使用它,获取更多的信息,从而做出更好更准确的决策。

在基于语料库评价出现的同时,一些翻译研究者试图通过设置评分评级标准来改进翻译质量评估。设计评分标准的主要目的在于纠正评估模式的还原主义倾向(即自下而上的方法),鼓励评估员采用自上而下的方法,更加全面地评价翻译作品的质量,关注更广泛的文本外因素(如文化适应性、读者反应、语境影响等)。而有了该类评估方式后,翻译质量评估的效率也能得到提升。这种新方法采用评分表的形式,其一般由三部分组成:(1)评分类别,与评估标准相对应;(2)评分段,旨在区分翻译质量/性能的连续体;(3)一系列渐进式的描述性语句或描述词,旨在捕捉每个级别翻译文本的突出特征。③ 在使用评分表时,评估人员首先需要阅读和分析特定的翻译文本,然后确定观察到的性能特征与每个评估标准的评分描述之间是否能够达到最佳匹配。为了检验该类评分标准的实用性,研究人员进行了大量的实证研究,经常将其与错误分析模型进行比较。通过具体计算和分析后,研究人员发现该类评分标准是一种可靠的方法;其结果与错误分析的结果在统计学意义上具有很强的关联性,而且评分者对不同语言的看法也有相对高的内部一致性。④ 虽然总体上此种评估方式是积极有效的,但它也存在一些潜在问题。研究人员在建立评分标准时,需要创建能精确描绘每个级别翻译特征的评分描述语言,并且对创建者的翻译能力、翻译经验等有

---

① Bowker L. Towards a Methodology for a Corpus-based Approach to Translation Evaluation[J]. Meta, 2001, 46 (2): 345-364.

② Jimenez-Crespo M A. A Corpus-based Error Typology: Towards A More Objective Approach to Measuring Quality in Localization[J]. Perspectives, 2001, 19 (4): 315-338.

③ Angelelli C V. Using a Rubric to Assess Translation Ability: Defining the Construct[M]//Angelelli C V, Jacobson H E. Testing and Assessment in Translation and Interpreting Studies. Amsterdam: John Benjamins, 2009: 13-47.

④ Colina S. Further Evidence for a Functionalist Approach to Translation Quality Evaluation[J]. Target, 2009, 21 (2): 215-244.

高要求,同时也需要其具有理论性和循证性的见解。① 许多研究者也对自上而下的评分方法表示担忧,认为它可能处于与错误分析模型相反的另一个极端。

鉴于之前描述的不同翻译评估模型代表了翻译质量评估的微观和宏观文本视角,沃丁顿(Waddington)呼吁将其结合起来,以发挥每种方法的最大潜力。② 这种混合方法后来⼜受到了大量研究,并在翻译批评中得到尝试。该类评估的方式流程主要如下:评估人员首先在翻译质量评估中分头行动,一方面用错误评估模型来检测翻译实践中错误的发生率,另一方面用评分表来全面评估翻译。然后,他们将两部分的分数结合起来,得出最终的等级或者分数。

虽然上述评估方法在翻译质量评估领域占主导地位,但一些应用语言学家也开发了基于项目的方法来评估翻译。通过比较上述的评估方式和该方式,研究者发现使用基于项目的评估方式可以产生最高水平的可靠性,并将薄弱学生与中级和高级学生区分开来。③ 尽管基于项目的评估方法具有较高的信度和区分度,但是其也有不足之处:首先,为了选择用于校准翻译测试的区分项目,必须确保学生样本具有能力连续从低到高的代表性,这种严格的样本类型一旦无法达到标准,就会限制该类方法的适用性;其次,该类翻译评估模式也无法处理大规模的翻译质量评估,而且该项目的耗时耗力也较大,并不具有普遍适用性。

在上述项目的基础上,几位中国学者介绍了一种替代方法,即比较判断或配对比较,从而达到评估翻译质量的目标。④ 其目的不是为了取代上述评估方法,而是为了补充完善翻译质量评估。该方法起源于心理物理分析,已被广泛用于教育评估。顾名思义,比较判断本质上是指专家比较两个类似的对象(在翻译评估的例子中则指同一源文本的两个译本),并对它们的相对质量做出二分决定(即好或不好),例如选择质量更高的翻译。因此,翻译质量是一个由个别专家感知的全局性、折衷性的属性。然后,对不同翻译对之间的连续比较,再对这样的二分法结果进行统计建模,以基于一定标准,为每个翻译产生质量估计。这些归一化的估计值可用于将每个翻译文本映射到一个从最差到最好的感知质量连续体上。这种方法背后的原理是人类与生俱来的能力,即相对判断:把一个对象与另一个对象进行比较。这样的方式比我们的绝

---

① Martinet Mateo R, Martinet Montero S, Guijarro A J M. The Modular Assessment Pack: A New Approach to Translation Quality Assessment at the Directorate General for Translation[J]. Perspectives, 2017, 25 (1): 18-48.

② Waddington C. Different Methods of Evaluating Student Translations: The Question of Validity[J]. Meta, 2001, 46 (2): 311-325.

③ Eyckmans J, Anckaert P, Segers W. The Perks of Norm-referenced Translation Evaluation[J]. Journal of Translators and Interprers, 2009, 3(2): 115-136.

④ Han C, Chen S-J, Fan Q. Rater-mediated Assessment of Translation and Interpretation: Comparative Judgement versus Analytic Rubric Scoring. Paper Presented at the 5th International Conference on Language Testing and Assessment[C]. Guangzhou, China: Guangdong University of Foreign Studies, 2019.

对判断——即单独判断一个对象的质量——更可靠和准确。① 此外,比较判断的有效性是基于累积的共识。尽管该类方式看来并不具备较高的可靠性,但是在某些情况下,该方法对翻译评估是有效的。其中原因主要在于以下几点:首先,不需要详细的评价标准就可以实现比较判断,避免了费尽心思地研究最适合不同语境的质量标准;其次,鉴于翻译研究学者普遍认为翻译质量是一种相对的而不是绝对的属性,这种共识为翻译的相对判断铺平了道路;最后,尽管质量的概念可能是主观的,但更多人相信,合格的、有经验的翻译专业人员和培训人员在评估翻译的相对适当性时,可以达到理想的一致程度。

国内的翻译质量评估模型也经过了一段时间的发展,其最早可以归结为吴新祥和李宏安提出的翻译等值观②,随后其他学者也建立了不同的翻译质量评估模型:如范守义等学者所提出的数学评估模型③;辜正坤等学者的最佳近似评估模型④;侯国金的语言标记等值评估模型⑤;司显柱的功能语言学评估模型⑥;何三宁的关联理论评估模型⑦。可以看出,国内学者对翻译质量评估及其模型进行了很好的研究,从不同的角度、用不同的方法对翻译质量评估进行了探索,力图取得突破。

"等值法"为构建翻译质量评估模型创造了一个有价值的新起点,但它不可避免地存在着自身的缺陷。所谓等值,是指"不但要求原作与译作有相同的信息、相同的思想、相同的形象、相同的意境、相同的情调,而且要求有相同的言语节奏、相同的言语风格、相同的言语韵味、相同的言语美学价值"。⑧ 事实上,翻译很难实现完全相同的再现,而翻译评价更不可能实现完全相同的对等性。此外,这种模式也忽视了翻译的艺术性。

"数学评估模型"是一种量化翻译质量的新方法,是规范翻译评估和大规模评估翻译质量的方法之一。然而,这种模式似乎过于理想化,也欠缺考虑翻译的审美价值。

"最佳近似评价模式"是从翻译标准体系的角度来分析翻译质量的。其切入点准确,同时这一翻译标准的建立对传统的"忠实"要求提出了挑战,在一定程度上揭示了翻译标准的规律,是评价翻译质量的另一个角度。其不足之处在于,标准的确定应从抽象到具体,但是在该模型中,具体的标准还是过于抽象,难以在实际情况中进行操作。

---

① Thurstone L L. A Law of Comparative Judgment[J]. Psychological Review, 1927(34):273-286.
② 吴新祥,李宏安.等值翻译初探[J]. 外语教学与研究,1984(3):2-10.
③ 范守义.模糊数学与译文评价[J]. 中国翻译,1987(4):2-9.
④ 辜正坤.翻译标准多元互补论[J]. 北京社会科学,1989(1):71-78.
⑤ 侯国金.浅论语用标记等效原则[J]. 山东外语教学,2005(1):17-20.
⑥ 司显柱.论功能语言学视角的翻译质量评估模式研究[J]. 外语教学,2004(4):45-50.
⑦ 何三宁."关联理论"视角下的翻译质量评估[J]. 南京师大学报(社会科学版),2010(1):155-160.
⑧ 吴新祥.等值论与译作定量定性分析[J]. 外语学刊,1985(1):15-28.

"功能语言学评价模式"考虑的参数比较全面,不仅有语言层面的句子、话语形式、话语功能等方面的分析,还有文本、文本类型及其功能角度的研究,具有一定的可信度,但该种模式还是"对文化环境的考虑相对欠缺"。① 此外,如果参数及其权重分配不当,难免会顾此失彼。例如,夏昭慧和曹合建二人在模型中采用了样本容量选择公式,在文体成分列表中,主要涉及词语、句法、修辞等。他们简单地将翻译定位在语言层面,忽视了翻译评价的其他方面,翻译质量评估的可靠性会大打折扣。②

"关联理论评价模式"追求的是原文和译文之间的意义趋同,以及语境效应的最大关联性。从这个意义上说,这种模式更适合于宏观评价,但在微观评价方面有所欠缺。

众所周知,翻译难,而要合理、准确地评价翻译也很难。此外,制定规范的评价标准、参数和模型,并结合案例进行实证分析更是难上加难。就国内现有的研究而言,确实出现了许多翻译评价模型的研究,也呈现了许多令人耳目一新的观点,但各种翻译质量评估模型的可操作性并不符合要求,还需进一步推进发展。长期以来,关于翻译的批评、评论一直是翻译批评界的重点。随着时代的发展,翻译活动不断深化,翻译质量的评估也在不断前进。

三、赖斯的翻译质量评估模型

在本章中,我们的主要目标在于评析航空航天类文学文本。在此基础上,我们拟使用赖斯的翻译质量评估模型来进行具体实践。赖斯的翻译质量评估模型在上文中已经提及,我们将在这里进行进一步阐释。

20 世纪 70 年代,德国著名学者赖斯在其著作《翻译批评:潜力与制约》中提出了一种基于原语语篇和目标语语篇功能关系的翻译批评模式。③ 赖斯对翻译理论家们提出的文本类型理论详细分析后,得出以下两条结论:第一,不可否认的是在选择翻译批评标准方面,文本类型起着首要的作用;第二,现存各种文本类型的分类都不尽如人意,因为它们在界定各种文本类型方面并没有表现出一致的原则,对文本作出区分的理由通常也是易变的、缺乏说服力的。④ 在这样的背景下,赖斯建立了一个可以容纳不同文本类型且适用于所有理论论域的翻译批评模式。

赖斯首先从语言功能来解决文本分类的争议,因为"文本的不同主要源于语言功能的不同"。⑤ 然而,对于语言功能的分类可以追溯到西塞罗时期,其中影响较大的

---

① 肖维青.翻译批评模式研究[M].上海:上海外语教育出版社,2009.
② 夏昭慧,曹合建.文体翻译对等的量化评估[J].湖南大学学报(社会科学版),2003(1):85-87.
③ 诺德.译有所为:功能翻译理论阐释[M].张美芳,王克非,主译.北京:外语教学与研究出版社,2005.
④ Reiss Katharina. Translation Criticism: The Potentials & Limitations[M]. Trans. by Erroll F Rhodes. Shanghai: Shanghai Foreign Language Education Press, 2004: 27-47.
⑤ 吴艾玲.莱斯的翻译类型学与文本类型翻译在中国[J].南京理工大学学报(社会科学版),2005(5):58-62.

是 1934 年德国心理学家、语言学家和符号学家布勒对于语言功能的分类。布勒将语言的功能归纳为三类,即陈述功能、表达功能和呼吁功能。[①] 赖斯沿用了布勒对语言功能的分类,将这三种语言功能与其相对应的语言维度、文本类型或各自使用的交际环境联系起来,将文本划分为以下三种类型:第一类文本是以内容为主的信息型文本,这类文本旨在交流、传递信息,文本语言的逻辑性以及文本内容的准确性很重要;第二类文本是以形式为主的表达型文本,这类文本是作者或文本发送者情感态度的表达,文本语言的表达形式和美学功能很重要;第三类文本是以诉请为主的呼吁型文本,这类文本旨在呼吁和感染读者采取某种行动,文本最后产生的效果十分重要。在描述完三种类型的文本的特征后,赖斯又提出了这三类文本的翻译方法。由此可以看出赖斯的文本类型理论中主要有三个要素,即文本类型、语言方面以及翻译方法,我们将语言方面细化后,其之间的对应关系见表 1 所列。[②]

表 1　赖斯翻译质量评估模型要素对应表

文本类型 (Text type)	信息型 (Informative)	表达型 (Expressive)	呼吁型 (Operative)
语言层面 (Language dimension)	逻辑 (Logic)	美学 (Aesthetics)	对话 (Dialogue)
语言功能 (Language function)	表现 (Representation)	表达 (Expressive)	劝导 (Persuasion)
翻译方法 (Translation method)	平实简洁 "Plain prose", explicitation as required	忠实贴切 "Identifying" method, adopt perspective of ST author	效果对等 "Adaptive", equivalent effect

除了以上这三种基于语言的主要功能分类的文本类型外,赖斯还注意到有一类以声音为媒介的文本,并将其增加为第四种类型。这类文本主要依赖一些非语言性的技术手段(如音响、画面等)来给观众传达听觉和视觉的意象。但是这类文本的划分在翻译界存在争议,学者们认为它和前面三类文本有重复的地方,因此,翻译理论家们往往忽略这类文本。

同时,本书接着讨论了各种非语言因素。例如,作者指出,译者和翻译批评家必须熟悉文本的内容,具有相应的知识,掌握相关领域的词汇。原作出版的年代和地点也是译者考虑的重要因素。如 18 世纪出版的小说和 20 世纪出版的小说,其译文必然有所不同。原文中许多有关地点的描写,可能充满了文化内涵,因此,译者必须熟悉有关的历史和文化知识。译者必须设法使译入语读者能够从译入语文化的角度理解原文。本书讨论的非语言因素还有原文中的情景(如某个人物说话时的情景)和与

[①] Munday Jeremy. Introducing Translation Studies[M]. London: Routledge, 2001.
[②] 同①。

说话者有关的各种因素，如受教育程度、生活的年代和说话的风格等。

在本章中，我们主要采取赖斯提出来的翻译质量评估模型来对航空航天类文学进行赏析，以期能更加全面客观地观察各类翻译现象。航空航天类文学属于科幻文学中的一类，作为一种通俗文学（或类型文学），由于其本身的"科学文学＋幻想文学"的特性，且常常兼具民族性和普世性的内容与题材，因而相较于其他文学类型（如主流文学或经典文学）更容易吸引读者的兴趣，其受众面更为广泛。① 其文本类型也具有更高的复杂性。在航空航天类文学中，其内容不仅涵盖航空航天类的技术信息，同时也具有较高的文学美性。因此，在赖斯的翻译质量评估模型中，其既属于有逻辑的信息型文本，也兼具重表达特质的表达型文本。译者在进行翻译时，需注意文本的这两大特征，在保证原文信息准确无误的前提下，尽可能地还原原文语言的美感。同时，在航空航天类文本中，也会出现信息类文字与表达类文字相分离的情况，其可能会出现在同一本书的不同情节，这也要求译者在翻译时需保持灵活的头脑，随时转换翻译方法，以达到最好的翻译效果。在本章的后两节中，我们将分别赏析航空航天类文学中英译汉和汉译英的一些翻译实践，以期能给予读者一些更切实的感受。

练 习 题

一、基础练习：请翻译以下句子

1. The night wore on, cold and clear, without further alarms, and the Moon rose slowly amid equatorial constellations that no human eye would ever see.②
2. In the caves, between spells of fitful dozing and fearful waiting, were being born the nightmares of generations yet to be.③
3. 我没见过黑夜，我没见过星星，我没见过春天、秋天和冬天。④
4. 我出生在刹车时代结束的时候，那时地球刚刚停止转动。⑤

二、拓展练习：请翻译以下段落

1. It whirled around, throwing its insanely daring tormentor against the wall of the cave. Yet whatever it did, it could not escape the rain of blows, inflicted on it by crude weapons wielded by clumsy but powerful hands. Its snarls ran the

---

① 熊兵. 中国科幻文学译介研究二十年（2000-2020）：回顾、反思与展望[J]. 外国语文研究，2022(6)，36-48.
② Clarke C Arthur. 2001: A Space Odyssey[M]. UK: Pearson Education Ltd & Penguin Books Ltd, 1968.
③ 同②.
④ 刘慈欣. 流浪地球[M]. 武汉：长江文艺出版社，2008.
⑤ 同④.

gamut from pain to alarm, from alarm to outright terror. The implacable hunter was now the victim, and was desperately trying to retreat.①

2. 比这景象更可怕的是发动机带来的酷热,户外气温高达七八十摄氏度,必须穿冷却服才能外出。在这样的气温下常常会有暴雨,而发动机光柱穿过乌云时的景象简直是一场噩梦!光柱蓝白色的强光在云中散射,变成无数种色彩组成的疯狂涌动的光晕,整个天空仿佛被白热的火山岩浆所覆盖。爷爷老糊涂了,有一次被酷热折磨得实在受不了,看到下大雨喜出望外,赤膊冲出门去,我们没来得及拦住他,外面雨点已被地球发动机超高温的等离子光柱烤热,把他身上烫脱了一层皮。②

三、思考题

1. 你认为翻译批评的研究意义是什么?
2. 翻译质量评估模型的优点和缺点分别是什么?
3. 你认为翻译批评未来的研究方向应该是什么?

第二节 航空航天文学翻译批评与赏析:英译汉

本节选取了不同航空航天类文本中的英译汉片段来进行具体赏析。同时,为了更好地评析翻译效果,本节也会就英汉之间的差异进行简要介绍,从而提出英译汉过程中需注意的一些事项。

一、英译汉的特点

清晰认识到英语和汉语之间的不同,有助于更好地进行翻译实践。英译汉的过程中,译者在读懂原文的基础上,更应明白汉语的用语习惯,以更好地达到目的语效果。

汉语属分析语,缺少形态变化,词序和助词是表达语法意义的主要手段。英汉互译时,往往要改变词性、转换词类才能通顺地表达原意。并且汉语有丰富多彩的助词,这是汉语的一大特点。助词又分为动态助词(如着、了、过)、结构助词(如的、地、得)和语气助词(如吗、呢、吧、啊、嘛、呀、哪、哇、呗、啦、罢了、似的等)。这些助词的作用,有一部分相当于英语的形态变化,有一部分却能左右结构、表达浓厚的感情色彩。③ 因此,我们在进行英译汉时,应多注意对助词的使用。如何将英文原文的意味

---

① Clarke C Arthur. 2001: A Space Odyssey[M]. UK: Pearson Education Ltd & Penguin Books Ltd, 1968.
② 刘慈欣. 流浪地球[M]. 武汉:长江文艺出版社,2008.
③ 连淑能. 略谈汉英语法特点[J]. 厦门大学学报(哲学社会科学版),1983(3):113-125.

全然翻译出,助词将会起到很重要的作用。同时,助词的使用也需注意准确性。在翻译的过程中,如果错误地使用了助词,就有可能导致原文的情感色彩、前后逻辑等出现问题。除此之外,英语经常使用定冠词和不定冠词。用不用冠词,在什么地方用冠词,用什么冠词,常常有正误之分或意思之别。因为汉语没有冠词,英译汉时可以省略,但有时一个冠词之差,意思大不相同。① 这就要求译者在进行翻译活动时要切实把握原文意思,注意英文中的冠词带来的意思差异。

从句子层面来看,汉语的主谓结构要复杂得多。主语不仅形式多样,而且可有可无:它可表示施事、受事,也可表示时间、地点;可用名词、动词,也可用形容词、数量词;句子可以没有主语,也可以省略主语,还可以变换主语并予以隐含。汉语的谓语也复杂多样:它可以是动词、名词或形容词;可以是一个动词,也可以是多个动词,还可以没有动词;它可以是一个单词,也可以是多个词组。汉语主谓结构具有很大的多样性、复杂性和灵活性,因而句式呈"流散型"。由于汉语重内在意念而不重外在形式,汉语的句型也就难以像英语那样以谓语动词为中心从形式上去划分。汉语是重语感、重变通的语言,组句的自由度很大,句子长长短短,不求形式齐整,而求意思通顺。许多表达形式灵活多变,往往靠约定俗成。汉语句式的多样化还表现在:有整句,也有大量的零句。整句有主谓结构,零句没有主谓结构,由词或词组构成。整句与零句混合交错,组成了流水句。② 因此,基于以上特点,英译汉的过程中,译者应将原句进行"拆分",注重汉语中"重意"的特点,可以使用丰富多样的句子形式来展现原文内涵。在保证忠实无误的基础上,译者也可以更好地通过调整句式的方式来展现语言特色。

二、英译汉文本赏析

航空航天类文本属于兼具信息型和表达型两种特点的文本类型。我们在运用赖斯的翻译质量评估模型对其翻译内容进行评估的同时,也可以多加考虑从不同角度对其进行赏析。与此同时,我们也要注意英译汉时语句的转换。在语言层面之外,我们也可考察翻译时的非语言因素对翻译造成的影响。

例1 英文原文:

The drought had lasted now for ten million years, and the reign of the terrible lizards had long since ended. Here on the Equator, in the continent which would one day be known as Africa, the battle for existence had reached a new climax of ferocity, and the victor was not yet in sight. In this barren and desiccated land, only the small or the swift or the fierce could

---

① 连淑能.略谈汉英语法特点[J].厦门大学学报(哲学社会科学版),1983(3):113-125.
② 连淑能.论英汉句法的基本特征[J].厦门大学学报(哲学社会科学版),1992(3):122-126.

flourish, or even hope to survive.①

中文译文：

　　这时，干旱已经持续了一千万年，可怕的恐龙也早已结束了主宰。在赤道此处，日后将以非洲之名而闻名的这块大陆上，求生之战的凶残，已沸腾到新的高点，胜出者则尚未见踪影。在这片干枯的不毛之地上，想要繁衍下去，或者起码有点存活下去的指望，就得要小，要快，要狠。②

　　这一段取自《2001：太空漫步》的部分，其主要功能是介绍故事的背景情况。首先从表现形式角度来看，英文原文仅由三个长句组成一段，其逻辑十分紧凑，并且动词在句子中的位置也比较灵活。翻译这样的片段时，译者需注意对语句的把握。我们可以看到，在这一段的译文中，中文段落是由许多零散的短句组成。这样的处理方式，不仅符合中文的语言习惯，同时也能通顺地表达原文的意思。同时，译者也善用四字词语，如"未见踪影""不毛之地"等词汇的使用也为文章增添了文采。针对"only the small or the swift or the fierce"这里的翻译时，译者很巧妙地调整语序，并改变了原词的词性，以动词的方式将其译出。"要小，要快，要狠"中的三个"要"字，就能生动形象地展现原文的情感色彩。但是，针对译文中"在赤道此处，日后将以非洲之名而闻名的这块大陆上，求生之战的凶残，已沸腾到新的高点，胜出者则尚未见踪影。"该句的处理还有待进一步讨论。这一句的语序与原文一致。但在中文里，一般不会在一个名词前添加一段冗长的定语来进行修饰。除此之外，译文忽视了原文的逻辑关系，忽略了对"yet"一词的翻译，导致译文无法完全体现原文的内部逻辑关系。在这里，如果译者将语序进行微调并将句子进行拆分以体现原文的逻辑关系，也许会达到更好的语言效果，如"在赤道的大陆，日后将以非洲一名而闻名。此处求生之战的凶残，已沸腾到新的高点，但胜出者却尚未见踪影。"从信息角度来看，译文将原文的意思较好地展现出来了。其中，对于"lizards"一词的翻译或许会引起争议。Lizard直接对应的中文词汇是"蜥蜴"，但是在这里，译者却选择了"恐龙"一词来进行阐释。其实，这不能算误译。在经过查询后，我们会发现恐龙的种类与名字是多种多样的，在中生代或中生代以前，有一类恐龙的名字就为"lizard"。根据原文的背景时间以及描述的环境情况，在此处翻译为恐龙是合理妥帖的。如果译者在这里直译为蜥蜴，也许读者也会困惑为何在那样的一个年代与环境中，小小的蜥蜴会成为主宰。

例 2　英文原文：

The ancients had, indeed, done better than they knew when they named this world after the lord of all the gods. If there was life down there, how long would it take even to locate it? And after that, how many centuries

---

　　① Clarke C Arthur. 2001: A Space Odyssey[M]. UK: Pearson Education Ltd & Penguin Books Ltd, 1968.
　　② 阿瑟•克拉克. 2001:太空漫步[M]. 郝明义,译. 上海:上海文艺出版社,2019.

before men could follow this first pioneer—in what kind of ship?①

中文译文：

的确，古人以"朱庇特"（Jupiter）这个众神之王的名字来为这个行星命名的时候，他们不知道自己做了多么棒的选择。就算那下面的确存在着生命，还要多久才能发现他们啊！之后，人类要想追随这第一个先驱者前进的话，还不知又要花上多少个世纪，要坐什么样的宇宙飞船啊！②

这一段的翻译中，一个很有意思的现象呈现了出来：原文的两个问号变成了译文中的两个感叹号。我们可以探讨这是否是一个妥帖的行为。如果我们将原文的两个问句进行直译，则其会分别译为"找到它需要多长时间？"和"在什么样的船上？"。不难发现，如果我们将该译文与例子中的译文进行对比，最大的区别就在于感情上。这里原文的问句并不是疑问句，而是反问句，其目的在于体现作者的感情色彩。译者充分抓取到了原文的这一特征，在翻译时灵活掌握句子特色，直接以中文的感叹句将其译出，不仅精巧，而且在情感上忠实于原文。此外，我们还会发现在这里的译文中，译者多加了一条信息，即"朱庇特"（Jupiter）。这个名字在原文是没有出现的，而译者选择在这里进行增译，其实也是为了使中文读者能够更好地理解原文。在西方文化中，罗马神话故事里的Jupiter是引领众神的王，这在以英语为母语的语言文化里是众所周知的一条信息。但是，在中国文化里，广大读者并不知晓这一文化点。因此，如果这里的翻译不明确这个众神之王的名字，读者也许不会理解原文所说的众神之王到底是谁。并且，作者除了在添加"朱庇特"这一名字时，还在后面添加了"（Jupiter）"这一英文信息。这其实是为了与前文呼应，更加明晰地告诉大家众神之王的名字与中文读者认知中的"木星（Jupiter）"是同一事物。在这里，我们能看到译者为了弥补中西方文化的信息差所做出的努力。

例3 英文原文：

Soft light sparkled onto the archways and tall columns of the opulent Caltech Athenaeum. It was an occasion like none other at JPL: the fiftieth anniversary of Explorer 1. On a January night in 2008, the institute celebrated the fateful day the first American satellite left Earth's atmosphere. Sadly, when making up the guest list for the anniversary, JPL forgot some important names. Five decades earlier, Barbara and Margie had sat in the control room and tracked the satellite as it flew through the sky, but in 2008, they were in their homes in Pasadena, just a few miles away from the celebration. They are two of the last people who remember JPL's control room that night, and

---

① Clarke C Arthur. 2001: A Space Odyssey[M]. UK: Pearson Education Ltd & Penguin Books Ltd, 1968.

② 阿瑟·克拉克. 2001：太空漫步[M]. 郝明义，译. 上海：上海文艺出版社，2019.

their work formed our first steps into space.①

中文译文：

柔和的光线照亮了加州理工雅典娜俱乐部（Caltech Athenaeum）的拱门和高柱，JPL 正在举行一场特殊的盛会：探险者 1 号 50 周年庆典。2008 年 1 月的一个晚上，JPL 决定以这种方式来纪念美国第一颗卫星离开大气层的重要日子。令人悲伤的是，拟定宴会宾客名单的时候，JPL 漏掉了几个重要的名字。50 年前，芭芭拉和玛姬曾坐在控制室里跟踪天空中那颗卫星的轨迹；但在 2008 年的这一天，她们却只能坐在帕萨迪纳的家里，虽然那里离庆典会场只有短短几英里。还记得 JPL 那间控制室的人已经不多了，芭芭拉和玛姬添列其间，她们的辛劳托起了我们迈向太空的第一步。②

本段是《让火箭起飞的女孩》终篇里的一部分，这里包含了作者丰富复杂的感情。虽然这本书讲述了许多航空航天类信息故事，但是最终也落脚在了"文学"一词上。在主角们为航空航天事业奋斗多年之后，身为女孩的她们却被人遗忘。这一段中的最后一句"They are two of the last people who remember JPL's control room that night, and their work formed our first steps into space."尤其能展现作者即将溢出的遗憾。"last"与"first"形成强烈对比。作为迈向太空极其重要的一步，为此奉献的这一群人却成了逐渐被遗忘的最后的人群。译者在处理这一句时，选择"不多了"来表达"last"的含义。这样的处理方式的确能够以一种符合中文使用习惯的方式表达原文意思，但却丧失了英文语句中的一些美感。译文里也丧失了原文的对比含义。这里是否能够有更好的处理方式，还值得进一步讨论。本段中"sadly"一词也被译者处理成了"令人悲伤的是"。这样的处理方式有欠妥帖。个人认为这里的 sadly 并不是在表达一种强烈的悲痛，更多的是作者为这群女孩们感受到的可惜与遗憾，所以这里的翻译如果变为"可惜的是"，也许会更妥帖。有时，相较于强烈的悲痛，一种淡淡萦绕在读者心中的遗憾之情反而更能打动人心。

练习题

一、基础练习：请翻译以下句子

1. The scene was so alien that for a moment it was almost meaningless to eyes accustomed to the colors and shapes of Earth.③

2. Far, far below lay an endless sea of mottled gold, scarred with parallel

---

① Holt Nathalia. Rise of the Rocket Girls[M]. New York: Little, Brown and Company, 2016.
② 娜塔莉娅·霍尔特. 让火箭起飞的女孩[M]. 阳曦，译. 北京：九州出版社，2022.
③ Clarke C Arthur. 2001: A Space Odyssey[M]. UK: Pearson Education Ltd & Penguin Books Ltd, 1968.

ridges that might have been the crests of gigantic waves.①

3. And that golden vista could not possibly have been an ocean, for it was still high in the Jovian atmosphere.②

二、拓展练习：请翻译以下段落

1. The night before the descent Maya couldn't sleep. Eventually she gave up trying, and pulled herself through the rooms and corridors, up to the hub. Every object was sharp-edged with sleeplessness and adrenaline, and every familiarity of the ship was countered or overwhelmed by some alteration, a lashed-down stack of boxes or a dead-end in a tube. It was as if they had already left the Ares. She looked around at it one last time, drained of emotion. Then she pulled herself through the tight locks, into the landing vehicle she had been assigned to. Might as well wait there. She climbed into her spacesuit, feeling, as she so often did when the real moment came, that she was only going through another simulation. She wondered if she would ever escape that feeling, if being on Mars would be enough to end it. It would be worth it just for that: to make her feel real for once! She settled into her chair.③

2. Late that afternoon Nadia tilted the nose of the dirigible down and circled into the wind, dropping until they were within ten meters of the ground and then releasing their anchor. The ship rose, jerked on its line, and settled downwind of the anchor, tugging at it like a fat kite. Nadia and Arkady twisted down the length of the gondola, to what Arkady called the bomb bay. Nadia lifted a windmill onto the bay's winch hook. The windmill was a little thing, a magnesium box with four vertical vanes on a rod projecting from its top. It weighed about five kilos. They closed the bay door on it, sucked out the air, and opened the bottom doors. Arkady operated the winch, looking through a low window to see what he was doing. The windmill dropped like a plumb and bumped onto hardened sand, on the southern flank of a small unnamed crater. He released the winch hook and reeled it back into the bay, and closed the bomb doors.④

---

① Clarke C Arthur. 2001: A Space Odyssey[M]. UK: Pearson Education Ltd & Penguin Books Ltd, 1968.
② 同①。
③ Robinson Stanley Kim. Red Mars[M]. New York: Bantam Books, 1993.
④ 同③。

第三节 航空航天文学翻译批评与赏析：汉译英

本节选取了不同航空航天类文本中的汉译英片段来进行具体赏析实践。与上一节类似，为了更好地评析翻译效果，本节也会就汉英之间的差异进行简要介绍，从而提出汉译英过程中需注意的一些事项。

一、汉译英的特点

汉译英的过程中，往往难点体现在译出的过程中。汉语作为中国译者的母语，在理解原文含义方面，译者一般不会出现任何问题，但是如何用原汁原味的英文体现原文全貌是一大难点。为了更好地企及这个目标，了解英语的语言特色十分重要。现代英语是从古英语发展过来的，仍然保留着综合语的某些特征，但也具有分析语的特点：有形态变化；词序比汉语灵活，但相对固定；虚词很多，用法也很灵活，但缺少像汉语那样灵活多变、生动传神的助词，等等。英语通过词形变化，改变词性，用这些词分别组句，可以表达一个几乎相同的意思。英语词序能够如此灵活倒置，形态变化和运用丰富的连接词是两个重要原因。英语有数量可观的关系词。关系词包括介词、连词、连接代词、关系代词、连接副词和关系副词。介词还可以和其他词构成合成介词和成语介词，组成形形色色的介词短语。英语常常通过这些关系词和形态变化，把词与词连接起来，造成长短句子，表达一定的逻辑关系。换言之，英语句子常用关系词和其他连接手段，注重句子形式，注重结构完整，这就是英语造句的主要方法。[①] 基于以上特征，汉译英的过程中，译者应更多使用关系词来连接句与句之间的关系。而且，英语是一类注重逻辑关系的语言，译者选择使用什么样的关系词至关重要。如若选择有误，就有可能引起译文逻辑不通，读者也无法理解作品内容。

英语句子有严谨的主谓结构。主语不可或缺，谓语是句子的中心，两者协调一致，提纲挈领，聚集各种关系网络。因此，英语句子主次分明，层次清楚，前呼后拥，严密规范，句式呈"聚集型"。英语句子复杂而不流散的另一重要原因是：句子成分之间或词语之间必须在人称、数、性和意义等方面保持协调一致的关系。这一原则包括三方面：(1)语法一致：即在语法形式上保持主语和谓语动词之间数与人称的一致，主语和表语之间数的一致，宾语和宾语补足语之间数的一致，人称代词、某些限定词和它们的照应对象之间的人称、数或性的一致，等等；(2)意义一致：即在意义上保持一致，如主语形式为复数，但意义为单数，谓语动词依意义采取单数形式，反之亦然；(3)就近原则：即谓语动词的人称和数往往和其最靠近的词语保持一致。英语句子成分之间这种协调一致的原则，使句子结构受到形态的约束，因而句式严谨、规范、刻板，缺

[①] 连淑能. 略谈汉英语法特点[J]. 厦门大学学报(哲学社会科学版), 1983(3): 113-125.

乏弹性。① 把握英语句子的特色,译者在翻译的过程中就不会"犯错"。

二、汉译英文本赏析

汉译英是一个考察译者英语能力的过程。以中文为母语的译者,对于理解原文不会有任何问题。然而,能否使用恰当的英语表达则考验了译者的能力。在对汉译英文本进行赏析时,我们仍然以是否忠实和表现手段是否恰当这两条为重要原则。同时,我们也需仔细观察译者在翻译过程中,对选词、语句的把握,感受其在翻译过程中细腻的程度。

例 4　中文原文:

地球发动机分为两大类,大一些的叫"山",小一些的叫"峰"。我们登上了"华北794号山"。登"山"比登"峰"花的时间长,因为"峰"是靠巨型电梯上下的,上"山"则要坐汽车沿盘"山"公路走。我们的汽车混在不见首尾的长车队中,沿着光滑的钢铁公路向上爬行。我们的左边是青色的金属峭壁,右边是万丈深渊。②

英文译文:

There were two major types of Earth Engines. The larger ones were dubbed "Mountains", while the smaller ones were called "Summits". We ascended North China Mountain 794. Let me tell you, it took a lot longer to scale a "Mountain" than to ascend a "Summit". The top of a Summit could be reached via a giant elevator, while you could only go up a Mountain in a car, snaking your way up a coiled road. Our bus weaved into the endless procession of other vehicles, following the smooth steel road up the outer side of the Mountain. To our left, there was only a blank face of azure metal; to our right, a yawning abyss.③

译者在处理本段时,十分精巧地使用了英文词汇来表达中文的含义。如"snaking""weaved into""endless"等词汇的使用,都能完美展现原文。译者用 snaking 一词来体现原文的"盘'山'公路",生动形象且具有极强的表达力。weave into 则展现了"混"的意蕴,如同织布一样混迹于车队之中,不但忠实于原文的意思,还加入了修辞手段,使译文更加生动,而 endless 则表达了不见首尾的无穷无尽。从这几个小小的词汇中,我们就能感受到译者对英文高超的掌握能力。在无数的英文词汇中,总能选择最合适的词汇,这是需要长时间的积累的。此外,我们能发现译者在翻译的时候添加了"let me tell you"。译者添加这一句,主要是为了拉近与读者的距离。流浪地球是一本从个人视角讲述故事的书,加上这一句后,作者与读者之间的

---

① 连淑能. 论英汉句法的基本特征[J]. 厦门大学学报(哲学社会科学版),1992(3):122-126.
② 刘慈欣. 流浪地球[M]. 武汉:长江文艺出版社,2008.
③ Liu Cixin. The Wandering Earth[M]. Trans. by Holger Nah. New York:Tor Trade, 2022.

距离进一步拉近。好像两个人的身份有了重合,看故事也就更加真切。本段还有另一个值得注意的地方,那就是对"山"和"峰"的处理。个人在阅读原文时,感受到作者这里形容的"峰"其实是指比山小一些的山丘,其强调的是"峰"较之于"山"的高度与形态,而不是山顶的意思。译文中,译者则选择了"summit"一词,这是否妥帖也值得进一步商榷。在牛津词典中,对"summit"的释义为"the highest point of sth, especially the top of a mountain",其强调的是"顶点"。因此,这里在忠实于原文信息这一条上或许能企及更高的高度。

例 5　中文原文：

"重元素聚变是一门很深的学问,现在给你们还讲不明白。你们只需要知道,地球发动机是人类建造的力量最大的机器,比如我们所在的华北794号,全功率运行时能向大地产生150亿吨的推力。"①

英文译文：

"Heavy element fusion is a very arcane field of study," she told me, "too difficult to understand at your age. You can content yourself to understand that the Earth Engines are the most powerful machines mankind has ever built. The one we are standing on, North China 794, operating at full power has the capability of exerting 15 billion tons of thrust on the Earth."②

本段的主要特色是具有极强的信息功能。里面涉及的一些信息类词汇如"重元素聚变""全功率运行""150亿吨",需要译者务必做到准确无误。在面对这类词汇的翻译时,译者往往需要查询相关资料,找到准确对应的英文,以避免在传达上出现失误。而在针对"很深的学问"这样具有语言个性的表达时,译者则要灵活处理,在对等层面上争取做到最佳。这里,如果译者使用直译的翻译手段,译文就会显得不伦不类。相反,译者选择了"arcane field"来进行表达,既能传达原文的意思,也十分符合英文读者的用语习惯,翻译得十分巧妙。

例 6　中文原文：

汪淼扭头一看,这声音是从正在燃烧的墨子发出来的,他的身体包含在一根高高的橘黄色火柱之中,皮肤在发皱和炭化,但双眼仍发出与吞噬他的火焰完全不同的光芒。他那已成为燃烧的炭杆的双手捧着一团正在飞散的绢灰,那是第一份万年历。汪淼自己也在燃烧,他举起双手,看到了两根火炬。③

英文译文：

Wang turned his head. The voice belonged to Mozi, who was already on fire. His body was encased within a column of tall, orange flame, and his skin

---

① 刘慈欣. 流浪地球[M]. 武汉:长江文艺出版社,2008.
② Liu Cixin. The Wandering Earth[M]. Trans. by Holger Nah. New York: Tor Trade, 2022.
③ 刘慈欣. 三体[M]. 重庆:重庆出版社,2008.

crinkled and turned into charcoal. But his two eyes still shone with a light that was distinct from the fire consuming him. His two hands, already burning pieces of charcoal, held up the cloud of swirling ashes that had once been his calendar.

Wang was burning up as well. He lifted his two hands and saw two torches.①

译者在对这一段进行翻译时,忠实于原文,完整展现了中文的每一个含义。我们可以看到译者首先理解了原文,然后运用关系代词、定语从句等语法手段,用英文将原文的意思串在一起。这一手段很好地贴近了读者,能够让读者在阅读时感到流畅通顺。这一段既没有很多的信息内容,也并不具有非常强的文学特色。在面对这样的文本内容时,译者就需要注意理解原文、忠实贴切和整合语句。然而,需要指出的是,这里的翻译有一个地方不太准确。原文里提及的是"万年历",译者翻译的却是"calendar"。虽然 calendar 也有万年历的意思,但是如果能够更明确一些,或许能够与上下文更加吻合,因此修改为"perpetual calendar"或许会是一个不错的选择。

练习题

一、基础练习:请翻译以下句子

1. "其实,人类把太阳同恐惧连在一起也只是这三四个世纪的事。这之前,人类是不怕太阳的,相反,太阳在他们眼中是庄严和壮美的。那时地球还在转动,人们每天都能看到日出和日落。他们对着初升的太阳欢呼,赞颂落日的美丽。"②

2. 自第一次氦闪后,随着重元素在太阳中心的反复聚集,太阳氦闪将在一段时间反复发生,这"一段时间"是相对于恒星演化来说的,其长度可能相当于上千个人类历史。③

3. 琳,你真的太聪明了,早在几年前,你就嗅出了知识界的政治风向,做出了一些超前的举动,比如你在教学中把大部分物理定律和参数都改了名字,欧姆定律改叫电阻定律,麦克斯韦方程改名成电磁方程,普朗克常数叫成了量子常数。④

二、拓展练习:请翻译以下段落

1. 沿着石阶,汪淼攀上了金字塔的顶部,看到了一处类似于古观星台的地方。平台的一角有一架数米高的天文望远镜,旁边还有几架较小型的。另一边是几台奇

---

① Liu Cixin. The Three-body Problem[M]. Trans. by Ken Liu. New York: A Tom Doherty Associates Book, 2018.
② 刘慈欣. 流浪地球[M]. 武汉:长江文艺出版社,2008.
③ 同②.
④ 刘慈欣. 三体[M]. 重庆:重庆出版社,2008.

形怪状的仪器,很像古中国的浑天仪。最引人注目的是平台中央的一个大铜球,直径两米左右,放置在一台复杂的机器上,由许多大小不同的齿轮托举着,缓缓转动。汪淼注意到,它的转动方向和速度在不停地变化。在机器下方有一个方坑,在里面昏暗的火光中,汪淼看到几个奴隶模样的人在推动着一个转盘,为上面的机器提供动力。[1]

2. "当然,记录员站在一个底部有滑轮的架子上,位置保持在球体中心。将模拟宇宙设定到现实宇宙的某一状态后,它其后的运转将准确地模拟出未来的宇宙状态,当然也能模拟出太阳的运行状态,那名记录员将其记录下来,就形成了一本准确的万年历,这是过去上百个文明梦寐以求的东西啊。你来得正好,模拟宇宙刚刚显示,一个长达四年的恒纪元将开始,汉武帝已根据我的预测发布了浸泡诏书,让我们等着日出吧。"[2]

---

[1] 刘慈欣. 三体[M]. 重庆:重庆出版社,2008.
[2] 同[1]。

第六章 航空航天文学翻译实践

本章为实践章节,请翻译以下文本,注意综合运用前几章所讲的技巧与方法。

第一节 航空航天文学翻译实践:英译汉

练习 1

Now the spinning wheels of light began to merge, and the spokes fused into luminous bars that slowly receded into the distance, rotating on their axes as they did so. They split into pairs and the resulting sets of lines started to oscillate across one another, slowly changing their angles of intersection. Fantastic, fleeting geometrical patterns flickered in and out of existence as the glowing grids meshed and unmeshed; and the man-apes watched, mesmerized captives of the shining crystal.

They could never guess that their minds were being probed, their bodies mapped, their reactions studied, their potentials evaluated. At first, the whole tribe remained half crouching in a motionless tableau, as if frozen into stone. Then the man-ape nearest to the slab suddenly came to life.[①]

练习 2

The stewardess came walking up the narrow corridor to the right of the closely spaced seats. There was a slight buoyancy about her steps, and her feet came away from the floor reluctantly as if entangled in glue. She was keeping to the bright yellow band of Velcro carpeting that ran the full length of the floor-and of the ceiling. The carpet, and the soles of her sandals, were covered with myriads of tiny hooks, so that they clung together like burrs. This trick of walking in free fall was

---

① Clarke C Arthur. 2001: A Space Odyssey[M]. UK: Pearson Education Ltd & Penguin Books Ltd, 1968.

immensely reassuring to disoriented passengers. ①

练习 3

Floyd was still musing over these thoughts when his helmet speaker suddenly emitted a piercing electronic shriek, like a hideously overloaded and distorted time signal. Involuntarily, he tried to block his ears with his spacesuited hands; then he recovered and groped frantically for the gain control of his receiver. While he was still fumbling, four more of the shrieks blasted out of the ether; then there was a merciful silence. ②

练习 4

Despite their relative youth, Poole and Bowman were veterans of a dozen space voyages, but now they felt like novices. They were attempting something for the first time; never before had any ship traveled at such speeds, or braved so intense a gravitational field. The slightest error in navigation at this critical point and Discovery would go speeding on toward the far limits of the Solar System, beyond any hope of rescue. The slow minutes dragged by Jupiter was now a vertical wall of phosphorescence stretching to infinity above them—and the ship was climbing straight up its glowing face. Though they knew that they were moving far too swiftly for even Jupiter's gravity to capture them, it was hard to believe that Discovery had not become a satellite of this monstrous world. ③

练习 5

Those responsible for piloting the Arespulled themselves to the control consoles and gave the orders to fire lateral control rockets. The Ares began to spin, stabilizing at four rpm. The colonists sank to the floors, and stood in a pseudogravity of 0.38 g, very close to what they would feel on Mars. Many man-years of tests had indicated that it would be a fairly healthy globe to live in, and so much healthier than weightlessness that rotating the ship had been deemed worth the trouble. And, Maya thought, it felt great. There was enough pull to make balance relatively easy, but hardly any feelingof pressure, of drag. It was the

---

① Clarke C Arthur. 2001: A Space Odyssey[M]. UK: Pearson Education Ltd & Penguin Books Ltd, 1968.
② 同①.
③ 同①.

perfect equivalent of their mood; they staggered down the halls to the big dining hall in Torus D, giddy and exhilarated, walking on air.①

练习 6

She hit the ground with both feet solid, nothing tricky about it, the g familiar from nine months in the Ares; and with the suit's weight, not that much different from walking on Earth, as far as she could remember. The sky was a pink shaded with sandy tans, a color richer and more subtle than any in the photos. "Look at the sky," Ann was saying, "look at the sky." Maya was chattering away, Sax and Vlad spun like rotating statues. Nadezhda Francine Cherneshevsky took a few more steps, felt her boots crunch the surface. It was salt-hardened sand a couple of centimeters thick, which cracked when you walked on it; the geologists called it duricrust or caliche. Her boot tracks were surrounded by small systems of radial fractures.②

练习 7

She walked along the shore of the new sea. It was at the foot of the Great Escarpment, in Tempe Terra, a lobe of ancient highlands extending into the north. Tempe had probably escaped the general stripping of the northern hemisphere by being roughly opposite the impact point of the Big Hit, which most areologists now agreed had struck near HradVallis, above Elysium. So, battered hills, overlooking an ice-covered sea. The rock looked like a red sea's surface in a wild cross chop; the ice looked like a prairie in the depths of winter. Native water, as Michel had said—there from the beginning, on the surface before. It was a hard thing to grasp. Her thoughts were scattered and confused, darting this way and that, all at the same time—it was like madness, but not. She knew the difference. The hum and keen of the wind did not speak to her in the tones of the MIT lecturer; she suffered no choking sensations when she tried to breathe. It was not like that. Rather her thinking was accelerated, fractured, unpredictable—like that flock of birds over the ice, zigzagging across the sky in a hard wind from the west. Ah the feel of that same wind against her body, shoving at it, the new thick air like a great animal's paw...③

---

① Robinson Stanley Kim. Red Mars[M]. New York: Bantam Books, 1993.
② 同①。
③ 同①。

练习 8

While the United States was thrown into World War Ⅱ in one devastating attack, China had been fighting since Japan invaded it in 1937. In a world descending into the chaos of war, Helen's mother had stood as the family's constant. By the 1940s, the Japanese held the fringes of China and were closing in. The family moved again and again, within and beyond China, trying to escape the escalating carnage. Helen's father, a general in Mao Tse-tung's Red Air Force, calculated his family's relocations based on military intelligence, but even those forewarnings couldn't keep them safe from the mounting destruction. When he moved the family to Hong Kong, his nerves calmed. The city was a safe haven under the protection of the powerful British Empire. The empire had never surrendered a colony; surely they would never submit to Japan.

These hopes shattered as the bombs rained down on the Pearl of the Orient. Helen's mother was at a neighbor's house. Trapped inside, she felt helpless to protect her children. When the thundering quieted, she raced home, shouting their names. She found Edwin and Helen clutching each other in the closet while her other two daughters crept out of their nearby hiding spot and hugged her close. Helen whispered softly, her voice faltering, "We thought you were gone."①

练习 9

Her fingers flying across her notebook, Helen noted when the first-stage Redstone fell off and the second stage of eleven Baby Sergeants fired for six seconds before falling back to Earth. Then the third stage of three Baby Sergeants fired, sending the delicate instruments even farther up through the atmosphere. The third stage fell away, and the final stage, despite being filled with sand instead of a satellite, was speeding farther than any man-made object had ever gone. Helen's eyes popped when she calculated that it had achieved a Mach 18 velocity and climbed 3,335 miles into the air, setting a new record for altitude. With the numbers confirmed, JPL erupted in excitement, while in Alabama von Braun literally danced with joy. The launch was more successful than they had dreamed possible. Helen gleefully slipped into Chinese exclamations as both elation and weariness washed over her. They had done it, even though there was no satellite.②

---

① Holt Nathalia. Rise of the Rocket Girls[M]. New York: Little, Brown and Company, 2016.
② 同①。

练习 10

The mission's goal, as defined by President Kennedy in 1961, was to "land a man on the moon and return him safely to Earth." To accomplish this, Apollo would send a three-man spacecraft into lunar orbit. Once it was circling the moon, a second spacecraft, the lunar module, would bring two astronauts down to the surface, leaving one man in the cone-shaped command module. All three astronauts would return to Earth in the command module, parachutes slowing their descent as they splashed down into the ocean. Despite their differences, Ranger and Apollo were faces of the same coin; they were both headed to the moon, one with men aboard and one without. While Project Apollo steadily advanced, NASA worried about their inability to get a lander on the moon. They needed Ranger to work.①

第二节　航空航天文学翻译实践：汉译英

练习 11

从我住的地方，可以看到几百台发动机喷出的等离子体光柱。你想象一个巨大的宫殿，有雅典卫城上的神殿那么大，殿中有无数根顶天立地的巨柱，每根柱子像一根巨大的日光灯管那样发出蓝白色的强光。而你，是那巨大宫殿地板上的一个细菌，这样，你就可以想象到我所在的世界是什么样子了。其实这样描述还不是太准确，是地球发动机产生的切线推力分量刹住了地球的自转，因此地球发动机的喷射必须有一定的角度，这样天空中的那些巨型光柱是倾斜的，我们是处在一个将要倾倒的巨殿中！南半球的人来到北半球后突然置身于这个环境中，有许多人会精神失常。②

练习 12

终于，我们看到了那令人胆寒的火焰，开始时只是天水连线上的一个亮点，很快增大，渐渐显示出了圆弧的形状。这时，我感到自己的喉咙被什么东西掐住了，恐惧使我窒息，脚下的甲板仿佛突然消失，我在向海的深渊坠下去，坠下去……和我一起下坠的还有灵儿，她那蛛丝般柔弱的小身躯紧贴着我颤抖着；还有其他孩子，其他的所有人，整个世界，都在下坠。这时我又想起了那个谜语，我曾问过哲学老师，那堵墙是什么颜色的，他说应该是黑色的。我觉得不对，我想象中的死亡之墙应该是雪亮

---

① Holt Nathalia. Rise of the Rocket Girls[M]. New York: Little, Brown and Company, 2016.
② 刘慈欣. 流浪地球[M]. 武汉：长江文艺出版社，2008.

的,这就是为什么那道等离子体墙让我想起了它。这个时代,死亡不再是黑色的,它是闪电的颜色,当那最后的闪电到来时,世界将在瞬间变成蒸汽。①

练习 13

人类的逃亡分为五步:第一步,用地球发动机使地球停止转动,使发动机喷口固定在地球运行的反方向;第二步,全功率开动地球发动机,使地球加速到逃逸速度,飞出太阳系;第三步,在外太空继续加速,飞向比邻星;第四步,在中途使地球重新自转,掉转发动机方向,开始减速;第五步,地球泊入比邻星轨道,成为这颗恒星的卫星。人们把这五步分别称为刹车时代、逃逸时代、流浪时代Ⅰ(加速)、流浪时代Ⅱ(减速)、新太阳时代。②

练习 14

我们的船继续航行,到了地球黑夜的部分,在这里,阳光和地球发动机的光柱都照不到,在大西洋清凉的海风中,我们这些孩子第一次看到了星空。天啊,那是怎样的景象啊,美得让我们心醉。小星老师一手搂着我们,一手指着星空,看,孩子们,那就是半人马座,那就是比邻星,那就是我们的新家!说完她哭了起来,我们也都跟着哭了,周围的水手和船长,这些铁打的汉子也流下了眼泪。所有的人都用泪眼探望着老师指的方向,星空在泪水中扭曲抖动,唯有那个星星是不动的,那是黑夜大海狂浪中远方陆地的灯塔,那是冰雪荒原中快要冻死的孤独旅人前方隐现的火光,那是我们心中的太阳,是人类在未来一百代人的苦海中唯一的希望和支撑……③

练习 15

我们就这样开始了地下的生活,像这样在地下 500 米处人口超过百万的城市遍布各个大陆。在这样的地下城中,我读完小学并升入中学。学校教育都集中在理工科上,艺术和哲学之类的教育已压缩到最少,人类没有这份闲心了。这是人类最忙的时代,每个人都有做不完的工作。很有意思的是,地球上所有的宗教在一夜之间消失得无影无踪。历史课还是有的,只是课本中前太阳时代的人类历史对我们就像伊甸园中的神话一样。④

练习 16

墨子对汪淼郑重地点点头,然后凑近他说,"知道吗,在你离开的三十六万两千年

---

① 刘慈欣. 流浪地球[M]. 武汉:长江文艺出版社,2008.
② 同①.
③ 同①.
④ 同①.

里,文明又重新启动了四次,在乱纪元和恒纪元的无规律交替中艰难地成长,最短的一次只走完了石器时代的一半,但139号文明创造了纪录,居然走到了蒸汽时代!"①

练习 17

"说得具体些,宇宙是一个悬浮于火海中的大空心球,球上有许多小洞和一个大洞,火海的光芒从这些洞中透进来,小洞是星星,大洞是太阳。"

"很有意思的一个模型,"汪淼看着铜球说,现在他大概能猜出那是什么了,"但其中有一个大漏洞:太阳升起和落下时,我们看到它与群星是相对运动的,而大球球壳上的所有洞孔的相对位置应该是固定的。"

"很对,所以我推出了经过修正的模型,宇宙之球是由两层球壳构成的,我们看到的天空是内层壳,外层球壳上有一个大洞,内层球壳上有大量小洞,那个外壳上的大洞透进的光在两层球壳之间的夹层反射和散射,使夹层间充满了亮光,这亮光从小洞中透进来,我们就看到了星星。"②

练习 18

墨子调出了游戏界面,将时间的流逝速度稍微调快了些。一轮红日升出地平线,大地上星罗棋布的湖泊开始解冻,这些湖泊原来封冻的冰面上落满了沙尘,与大地融为一体,现在渐渐变成一个个晶莹闪亮的镜面,仿佛大地睁开了无数只眼睛。在这高处,浸泡的具体细节看不清楚,只能看到湖边的人渐渐多了起来,像春天涌出洞穴的蚁群。世界再一次复活了。③

练习 19

墨子从一只木箱中拿出了一块黑色圆片,"加上这片烟熏的滤镜。"说着将它插到望远镜的目镜前。汪淼将望远镜对准已升到半空的太阳,不由赞叹墨子的想象力:太阳看上去确实像一个通向无边火海的孔洞,是一个更大存在的一小部分。但进一步细看时,他发现,这个太阳与自己现实经验中的那个有些不同,它有一颗很小的核心,如果将太阳看成一只眸子,这个日核就像瞳孔。日核虽小,但明亮而致密,包裹它的外层则显得有些缺少实在感,飘忽不定,很像是气态的。而穿过那厚厚的外层能看到内部日核,也说明外层是处于透明或半透明状态的,它发出的光芒,更多的可能是日核光芒的散射。④

---

① 刘慈欣. 三体[M]. 重庆:重庆出版社,2008.
② 同①。
③ 同①。
④ 同①。

练习 20

那片银白色的曙光以超乎寻常的速度扩展变亮,仿佛即将升起的太阳要弥补失去的时间。很快,曙光已弥漫了半个天空,以至太阳还未升起,大地已同往日的白昼一样明亮。汪淼向曙光出现的远方看去,发现地平线发出刺眼的强光,并向上弯曲拱起,成一个横贯视野的完美弧形,他很快看出那不是地平线,是日轮的边缘,正在升起的是一颗硕大无比的太阳!眼睛适应了这强光后,地平线仍在原位显现出来,汪淼看到一缕缕黑色的东西在天边升起,在日轮明亮的背景上格外清晰,那是远方燃烧产生的烟雾。金字塔下面,一匹快马从日出方向飞驰而来,扬起的尘埃在大地上划出一道清晰的灰线,人群为其让开了一条路,汪淼听到马上的人在声嘶力竭地大喊:"脱水!脱水!!"[①]

---

① 刘慈欣. 三体[M]. 重庆:重庆出版社,2008.

练习题及翻译实践的参考答案

第一章
第一节
练习题

一、基础练习：请翻译以下句子

1. She had just finished setting it to rights, and was shaking her duster from the window, when she saw the two men again.①

她刚整理好自己的房间，在窗户前抖抹布的时候又看见了这两个男人。②

2. They were standing now by the car and, Miss Clare was glad to see, they were doing their best to wipe the mud from their shoes on the grass.③

他们站在车旁，令克莱尔女士欣慰的是他们正在草上努力地除去鞋上的泥巴。④

3. 悖时砍脑壳的！⑤

"Damned low-life! You're headed for the executioner!"⑥

4. 横顺人是"牛肉炒韭菜，各人心里爱。"⑦

Anyway, as the saying goes, "people eat what they like, even beef with chives."⑧

二、拓展练习：请翻译以下段落

1. "They'd got a new-fangled thing-brief-case, ain't it? —in the back. Two strangers, poking about here with a brief-case and a lot of mud on their shoes," she, mused. "Makes you think, don't it? Might be Ag. men, of course. But you mark my words, Miss Read, they was up to a bit of no good!"⑨

"他们车后面还放了个新玩意儿，那不是个公文包吗？两个陌生人夹个公文包来到这里，脚上还沾满了泥，"她心里嘀咕着，"让人捉摸不透，不是吗？当然也可能是什

---

① 宋楠.语境理论背景下文学翻译策略的选择和应用:以英语小说《乡村风暴》翻译实践为例[J].牡丹江大学学报,2015,24(5):135-138+147.

② 同①。

③ 同①。

④ 同①。

⑤ 沈从文.沈从文全集:第8卷[M].太原:北岳文艺出版社,2002.

⑥ Kinkley Jeffrey C. Border Town[M]. New York: HarperCollins Publishers, 2009.

⑦ 同⑤。

⑧ 同⑥。

⑨ 同①。

么检察官之类的。但是瑞德小姐,你记住,他们来这里,绝不是什么好事。"①

2. 车是车路,马是马路,各有规矩。想爸爸作主,请媒人正正经经来说是车路;要自己作主,站到对溪高崖竹林里为你唱三年六个月的歌是马路。②

Chariots have to move like chariots, and horse-men like horsemen, according to the rules. If his father was going to take charge of this, he had to have a matchmaker do it according to custom—that's how chariots move; if he wanted to take charge of it himself, he had to go up into the bamboo grove on the bluffs across the creek and sing for you, three years and six months—that's the horseman's move.③

第二节
练习题

一、基础练习:请翻译以下句子

1. Inside our own galaxy, there are roughly 100 million smaller black holes: holes that typically are between about three and thirty times as heavy as the Sun.④

在星系中,大约存在着1亿个小质量黑洞。它们的质量大多是3～30倍太阳质量。⑤

2. We know this not because we've seen evidence for all these, but because astronomers have made a census of heavy stars that will become black holes when they exhaust their nuclear fuel.⑥

我们并没有看到所有这些黑洞的存在证据,而是通过天文学家们对大质量恒星的普查知道了这一点,这些恒星未来都能够变成黑洞。⑦

3. Barbara put on a modest shirtdress, the hem skimming her calves.⑧

芭芭拉穿上了一条朴素的衬衫裙,裙摆轻轻拍打着她的小腿。⑨

4. She buttoned the collar down conservatively but then, as a nod to her femininity, cinched the belt at her waist, showing off her slim figure.⑩

---

① 宋楠. 语境理论背景下文学翻译策略的选择和应用:以英语小说《乡村风暴》翻译实践为例[J]. 牡丹江大学学报,2015,24(5):135-138+147.
② 沈从文. 沈从文全集:第8卷[M]. 太原:北岳文艺出版社,2002.
③ Kinkley Jeffrey C. Border Town[M]. New York: Harper Collins Publishers, 2009.
④ Kip T. The Science of Interstellar[M]. New York: W. W. Norton & Company, 2014.
⑤ 基普·索恩. 星际穿越[M]. 苟利军,王岚,李然,等,译. 杭州:浙江人民出版社,2015.
⑥ 同④。
⑦ 同⑤。
⑧ Holt N. Rise of the Rocket Girls[M]. New York: Little, Brown and Company, 2016.
⑨ 娜塔莉娅·霍尔特. 让火箭起飞的女孩[M]. 阳曦,译. 北京:九州出版社,2022.
⑩ 同⑧。

她保守地扣紧了领口,但纤细的腰身完美地衬托出了她柔美的身形。①

二、拓展练习:请翻译以下段落

1. Before the voting commenced, however, Barbara's co-workers hoisted her atop a convertible and drove around the lab. With the wind blowing through her hair, she smiled and waved. She felt a little silly and laughed nervously. Barbara might not be the prettiest girl in the lab, but she was sociable and easy to work with. All the computers were rooting for her. She imagined the director of the institute crowning her at the summer dance.②

可是投票还没开始,同事们就簇拥着芭芭拉登上一辆敞篷汽车,带她出去兜风了。风儿在她的发丝间穿梭,她一边笑一边挥手。这样的举动让她觉得自己有点傻,于是她笑得更大声了。芭芭拉或许不是实验室里最漂亮的姑娘,但她待人热情,易于共事。所有计算员都是她的铁杆。她不禁开始想象实验室主任在夏夜舞会上为她戴上桂冠的情景。③

2. Yet she didn't dwell on her possible victory, since it was merely a lighthearted affair. No one took it too seriously. Barbara, representing the computing section, was competing against Lois Labee, a chemist, and Margaret Anderson, who worked in the research design division. They were all young, beautiful, and very good at their demanding jobs. As odd as it seems by today's standards, the beauty contest was a result of JPL's progressive hiring practices. As the bouquets were handed out and an attractive woman crowned the winner, the competition was unintentionally highlighting the presence of educated young women working at JPL. After all, other laboratories would have found it impossible to hold such a contest in the 1950s; they simply didn't hire enough women.④

但触手可及的胜利没有冲昏她的头脑,因为这场比赛不过是个玩闹的小把戏,谁也不会拿它太当回事。芭芭拉是计算部门的代表,她的对手是化学家洛伊丝·拉比和研究设计部的玛格丽特·安德森。这几个漂亮姑娘都很年轻,干起活儿来也得心应手。虽然以今天的标准来看有些奇怪,但这场选美大赛却是JPL先进的人才策略带来的结果。漂亮的姑娘手持花束戴上胜利的冠冕,这样的比赛不经意地让JPL受过良好教育的年轻女员工成为人们关注的焦点。归根结底,20世纪50年代的其他实验室就算想搞这样的比赛也有心无力:他们根本没有那么多女雇员。⑤

---

① 娜塔莉娅·霍尔特. 让火箭起飞的女孩[M]. 阳曦,译. 北京:九州出版社,2022.
② Holt N. Rise of the Rocket Girls[M]. New York: Little, Brown and Company, 2016.
③ 同①。
④ 同②。
⑤ 同①。

第三节
练习题

一、基础练习：请翻译以下句子

1. The mission's goal, as defined by President Kennedy in 1961, was to "land a man on the moon and return him safely to Earth."①

正如肯尼迪总统在1961年说过的那样，这个任务的目标是"将人类送上月球，并把他安全带回地球"。②

2. Once it was circling the moon, a second spacecraft, the lunar module, would bring two astronauts down to the surface, leaving one man in the cone-shaped command module.③

接下来，第二艘飞船——登月舱——将带着两位宇航员登上月球表面，另一位宇航员则留在圆锥形的指令舱里。④

3. This really superb piece of abstract thinking led Moon-Watcher, after only three or four minutes, to a deduction which he immediately put to the test.⑤

只经过三四分钟之后，这段无与伦比的抽象思索，帮望月者导出一个他立即付诸测试的结论。⑥

4. They could never guess that their minds were being probed, their bodies mapped, their reactions studied, their potentials evaluated.⑦

他们做梦也不会想到的是：在这段时间，他们的心智正在被探测，体态正在被记录，反应正在被研究，潜能正在被评估。⑧

二、拓展练习：请翻译以下段落

It whirled around, throwing its insanely daring tormentor against the wall of the cave. Yet whatever it did, it could not escape the rain of blows, inflicted on it by crude weapons wielded by clumsy but powerful hands. Its snarls ran the gamut from pain to alarm, from alarm to outright terror. The implacable hunter was now the victim, and was desperately trying to retreat.⑨

它打了个转，把这个胆大包天的加害者甩上了洞壁。然而，不论它采取什么行动，都没法躲开四面如雨而下的攻击——一双双笨拙却有力的手，舞动着一些粗糙武

---

① Holt N. Rise of the Rocket Girls[M]. New York: Little, Brown and Company, 2016.
② 娜塔莉娅·霍尔特. 让火箭起飞的女孩[M]. 阳曦，译. 北京：九州出版社，2022.
③ 同①。
④ 同①。
⑤ Clarke C A. 2001: A Space Odyssey[M]. UK: Pearson Education Ltd & Penguin Books Ltd, 1968.
⑥ 阿瑟·克拉克. 2001：太空漫步[M]. 郝明义，译. 上海：上海文艺出版社，2019.
⑦ 同⑤。
⑧ 同⑥。
⑨ 同⑤。

器而进行的攻击。它嘶吼的声音,从疼痛转为惊慌,从惊慌转为彻底的恐惧。现在,这个横行无阻的狩猎者,转而成了受害者,一心一意只想撤退。①

三、阅读实践

(略)

第二章
第一节
练习题

一、基础练习:请翻译以下句子

1. "A scandal if that land is taken for building!" he said, chopping up a piece of chocolate cake viciously. "More larks there to the square yard than anywhere else in England!"②

"如果那块地被用来建盖住宅小区的话,会激起民愤的!"他说着,狠狠地切了一块巧克力蛋糕。"那块方形地上空,云雀比英格兰任何一个地方都多。"③

2. "And if the old Tartar finds out, it is all one to me!" he added sturdily, tucking it behind the sack which shrouded it.④

"如果那个老悍妇发现了,也没什么!"他坚定地说,把煤油罐塞回去,再用一个布袋子把它盖起来。⑤

3. "Then I'll come definitely every Orchestra night," I promised. "I should have thought of it before. It's the least a godmother can do."⑥

"那么每晚乐队演奏的时候,我都过来帮你照看孩子,"我承诺道。"我应该早就想到的,这是一个教母该做的一点小事。"⑦

4. Hundred Acre Field and its spacious neighbors were among the more fruitful parts of Mr. Miller's farm.⑧

百亩田园和它广阔的临近地区是米勒先生农场最肥沃的地区之一。⑨

二、拓展练习:请翻译以下段落

黄昏以前老道士用红绿纸剪了一些花朵,用黄泥做了一些烛台。天断黑后,棺木前小桌上点起黄色九品蜡,燃了香,棺木周围也点了小蜡烛,老道士披上那件蓝麻布

---

① 阿瑟·克拉克. 2001:太空漫步[M]. 郝明义,译. 上海:上海文艺出版社,2019.
② 宋楠. 语境理论背景下文学翻译策略的选择和应用:以英语小说《乡村风暴》翻译实践为例[J]. 牡丹江大学学报,2015,24(5):135-138+147.
③ 同②。
④ 同②。
⑤ 同②。
⑥ 同②。
⑦ 同②。
⑧ 同②。
⑨ 同②。

道袍,开始了丧事中的绕棺仪式。老道士在前面拿着小小纸幡引路,孝子第二,马兵殿后,绕着那具寂寞棺木慢慢转着圈子。①

Before dusk fell, the old priest cut out some flower shapes from red and green paper and fashioned candlesticks from yellow mud. When it was dark, a yellow candle was lit on the small table in front of the casket. There was incense, and other little candles were lit all around the coffin as the old Daoist priest put on his blue hempen gown and began the funeral rite of circling the coffin. The old priest went in front, carrying a paper streamer to lead the way, followed by the filially pious daughter and the horseman in the rear, slowly going in a circle around the lonely casket.②

第二节
练习题

一、基础练习:请翻译以下句子

1. I was descending a steep, cobbled, excavated road between banked-up footways, perhaps six feet high, upon which, in a monotonous series, opened the living room doors of rows of dark, low cottages.

当时我正沿着一条用鹅卵石铺成的陡峭小路往下走。这条路是从人行道中间开掘出来的,两侧的壁坡约有六英尺高,上面有一排排式样单调的又黑又矮的小屋,屋门洞开。

2. The perspective of squat blue slate roofs and clustering chimneys drifted downward towards the irregular open space before the colliery—a space covered with coaly, wheel-scarred mud, with a patch of weedy dump to the left and the colliery gates to the right.

蓝色的低矮的石板瓦屋顶和林立的大烟囱所形成的视点不断向下延伸,逐渐与煤矿前那片不规则的空地相连。空地上满是煤灰和车辙的泥印。左边是一片杂草丛生的垃圾堆,右边是煤矿的大门。

二、拓展练习:请翻译以下段落

Beyond, the High Street with shops resumed again in good earnest and went on, and the lines of the steam-tramway that started out from before my feet, and were here shining and acutely visible with reflected skylight and here lost in a shadow, took up for one acute moment the greasy yellow irradiation of a newly lit gas lamp as they vanished round the bend.

远处,商店林立的大街重新开张,继续热闹起来,蒸汽机车的行驶线路就从我脚

---

① 沈从文.沈从文全集:第8卷[M].太原:北岳文艺出版社,2002.
② Kinkley Jeffrey C. Border Town[M]. New York: HarperCollins Publishers, 2009.

下延伸开去,可以清晰地看见反射的天光在闪烁,然后消失在阴影里,有那么一会儿,一盏新点燃的煤气灯发出油腻的黄色光亮,再逐渐消失在拐弯处。

第三节
练习题

一、基础练习:请翻译以下句子

1. 老刀从来没有见过这样的景象。太阳缓缓升起,天边是深远而纯净的蓝,蓝色下沿是橙黄色,有斜向上的条状薄云。

Lao Dao had never seen such a sight. The sun rose gradually. The sky was a deep and pure azure, with an orange fringe at the horizon, decorated with slanted, thin wisps of cloud.

2. 太阳被一处屋檐遮住,屋檐显得异常黑,屋檐背后明亮夺目。

The eaves of a nearby building blocked the sun, and the eaves appeared especially dark while the background was dazzlingly bright.

3. 太阳升起时,天的蓝色变浅了,但是更宁静透彻。

As the sun continued to rise, the blue of the sky faded a little, but seemed even more tranquil and clear.

4. 他想抓住那道褪去的金色。蓝天中能看见树枝的剪影。他的心狂跳不已。他从来不知道太阳升起竟然如此动人。

He wanted to catch a trace of that fading golden color. Silhouettes of waving tree branches broke up the sky. His heart leapt wildly. He had never imagined that a sunrise could be so moving.

二、拓展练习:请翻译以下段落

他跑了一段路,停下来,冷静了。他站在街道中央。路的两旁是高大树木和大片草坪。他环视四周,目力所及,远远近近都没有一座高楼。他迷惑了,不确定自己是不是真的到了第一空间。他能看见两排粗壮的银杏。

After a while, he slowed down and calmed himself. He was standing in the middle of the street, lined on both sides with tall trees and wide lawns. He looked around, and he couldn't see any buildings at all. Confused, he wondered if he had really reached First Space. He pondered the two rows of sturdy gingkoes.

第三章
第一节
练习题

一、基础练习:请翻译以下句子

1. The old dam we interrupted at her feeding on the Green Hill swept Balmy with a cold glance, kicked up her heels, and trotted toward the centre of the herd flinging over her shoulder at her spindly-legged foal a command to follow. But the

foal never moved.

那匹在绿丘上吃草的母斑马因为受到打扰,所以冷冷地扫了"小古怪"一眼,然后扬起后蹄,缓缓走向斑马群,还扭头对它四肢颤巍巍的幼崽下了道指令,要它跟上。但那匹小斑马却一动不动。

2. The wind in the wires is like the tearing of soft silk under the blended drone of engine and propeller.

线路中的风声就像柔软的丝绸正被引擎和螺旋桨协力撕碎。

3. Time and distance together slip smoothly past the tips of my wings without sound, without return, as I peer downward over the night-shadowed hollows of the Rift Valley and wonder if Woody, the lost pilot, could be there, a small human pinpoint of hope and of hopelessness listening to the low, unconcerned song of the Avian—flying elsewhere.

时间与距离在我的机翼下无声滑过,永不复返。我向下俯瞰大裂谷的暗影,心想,那个失踪的飞行员伍迪是否会在那里。是否正带着渺茫的希望与无望,倾听飞机吟唱着低沉而冷漠的曲调,飞向他方。

4. Everywhere on earth that day, in the ears of every one who breathed, there had been the same humming in the air, the same rush of green vapors, the crepitation, the streaming down of shooting stars.

那天,在地球上的每一个地方,每个呼吸着的人都能听到空气中传来同一种喃喃声,感觉到绿色气体的冲击,看见星星碰撞时迸发的火光与气流的扇动。

二、拓展练习:请翻译以下段落

1. The Hindoo had stayed his morning's work in the fields to stare and marvel and fall, the blue-clothed Chinaman fell headforemost athwart his midday bowl of rice, the Japanese merchant came out from some chaffering in his office amazed and presently lay there before his door, the evening gazers by the Golden Gates were overtaken as they waited for the rising of the great star.

印度人在地里干着早上的农活,他们瞪着眼睛惊奇地望着天空,然后倒在地上;穿着蓝色衣服的中国人埋头于午餐的饭碗上;刚从办公室出来的日本商人感到十分惊讶,这时也躺在了办公室的门前;金库的守夜人正在等着看巨大的恒星升空,却被什么东西击倒了。

2. As I flew, my hunch became conviction. Nothing in the world, I thought, could have looked so much like reflecting water as the wings of Woody's plane. I remembered how bright those wings had been when last I saw them, freshly painted to shine like silver or stainless steel.

当我飞行的时候,我的直觉愈发坚定。我觉得世界上再没有别的东西会比伍迪飞机的机翼更像反光的水面了。我记得上次看见它们时,那机翼是多么明亮,刚刷过新漆,亮得像白银或不锈钢。

第二节
练习题

一、基础练习:请翻译以下句子

1. This convoluted critical orbit is a close analog of the trajectories of temporarily trapped light rays inside Gargantua's shell of fire. ①

这一错综复杂的临界轨道非常类似于光线被暂时困在卡冈都亚火壳之内时的轨道。②

2. Like those light rays, the Endurance is temporarily trapped when on its critical orbit. ③

与这些光线一样,"永恒"号在其临界轨道上被临时捕获。④

3. Unlike the light rays, the Endurance has a control system and rockets, so its launch off the critical orbit is in Brand's and Case's hands. ⑤

与光线不同的是,"永恒"号拥有控制系统以及火箭,因此布兰德和凯斯能够通过发射火箭让它脱离临界轨道。⑥

4. And because of the orbit's convoluted three-dimensional structure, the launch can be in any direction they wish. ⑦

并且,正是由于复杂的三维轨道结构,发射能够如他们所希望的那样朝向任何一个方向。⑧

二、拓展练习:请翻译以下段落

1. The Endurance heads toward Gargantua with a certain amount of energy, which like its angular momentum remains constant along its trajectory. This energy consists of three parts: the Endurance's gravitational energy, which gets more and more negative as the Endurance plunges toward Gargantua; its centrifugal energy (its energy of circumferential motion around Gargantua), which increases as the Endurance plunges because the circumferential motion is speeding up; and its radial kinetic energy (its energy of motion toward Gargantua). ⑨

"永恒"号向卡冈都亚运动时会带有一定的能量。与角动量一样,能量沿着它的

---

① Kip T. The Science of Interstellar[M]. New York: W. W. Norton & Company, 2014.
② 基普·索恩. 星际穿越[M]. 苟利军,王岚,李然,等译. 杭州:浙江人民出版社,2015.
③ 同①。
④ 同②。
⑤ 同①。
⑥ 同②。
⑦ 同①。
⑧ 同②。
⑨ 同①。

轨迹也是不变的。这一能量由三部分组成:"永恒"号的引力能——随着"永恒"号驶向卡冈都亚的距离越近,引力能负值越大;它的离心能(围绕卡冈都亚做圆周运动的能量)——随着"永恒"号驶向黑洞的距离越近,圆周速度越快,值越大;以及,它的径向动能(朝向卡冈都亚运动的能量)。①

2. Our discussion of Gargantua's environs has taken us from the physics of planets (tidal deformation, tsunamis, tidal bores,), through Gargantua's vibrations and the search for organic signs of life, to engineering issues (the Endurance's robust design and its damaging explosion). As much as I enjoy these topics—and I've done research or textbook writing on most of them— they are not my greatest passion. My passion is extreme physics; physics at the edge of human knowledge and just beyond. That's where I take us next.②

我们对于黑洞卡冈都亚周边环境的讨论将我们从行星物理(潮汐力形变、海啸、涌潮等)带到了卡冈都亚的振动以及对有机生命存活迹象的搜寻,又带到了工程问题("永恒"号的坚固设计和它破坏性的爆炸)。虽然我非常喜欢这些话题,并且研究过它们中的绝大部分并在此基础上写过教材,但它们并不是我现在最热衷的部分。我现在的热情在极端物理上,即那些处于人类认知边缘以及超越我们认知范围的物理上。这正是我将要谈到的。③

第四章
第一节
练习题

一、基础练习:请翻译以下句子

1. 一天叶文洁值夜班。这是最孤寂的时刻。在静静的午夜,宇宙向它的聆听者展示着广漠的荒凉。

One night, Ye was working the night shift. This was the loneliest time. In the deep silence of midnight, the universe revealed itself to its listeners as a vast desolation.

2. 叶文洁最不愿意看的,就是显示器上缓缓移动的那条曲线,那是红岸接收到的宇宙电波的波形。无意义的噪声。

What Ye disliked most was seeing the waves that slowly crawled across the display, a visual record of the meaningless noise Red Coast picked up from space.

3. 叶文洁感到这条无限长的曲线就是宇宙的抽象,一头连着无限的过去,另一头连着无限的未来,中间只有无规律无生命的随机起伏。

---

① 基普·索恩. 星际穿越[M]. 苟利军,王岚,李然,等译. 杭州:浙江人民出版社,2015.
② Kip T. The Science of Interstellar[M]. New York: W. W. Norton & Company, 2014.
③ 同①.

Ye felt this interminable wave was an abstract view of the universe: one end connected to the endless past, the other to the endless future, and in the middle only the ups and downs of random chance—without life, without pattern.

4. 一个个高低错落的波峰就像一粒粒大小不等的沙子,整条曲线就像是所有沙粒捧成行形成的一堆沙漠,荒凉寂寥,长得更令人无法忍受。你可以沿着它向前向后走无限远,但永远找不到归宿。

The peaks and valleys at different heights like uneven grains of sand, the whole curve like a one-dimensional desert made of all the grains of sand lined up in a row, lonely, desolate, so long that it was intolerable. You could follow it and go forward or backward as long as you liked, but you'd neverfind the end.

二、拓展练习:请翻译以下段落

1. 地球的变轨加速就这样年复一年地进行着。每当地球向远日点升去时,人们的心也随着地球与太阳距离的日益拉长而放松;而当它在新的一年向太阳跌去时,人们的心一天天紧缩起来。每次到达近日点,社会上就谣言四起,说太阳氦闪就要在这时发生了;直到地球再次升向远日点,人们的恐惧才随着天空中渐渐变小的太阳平息下来,但又在酝酿着下一次的恐惧……①

The Earth's orbitally-assisted acceleration carried on like this, year after year. Every time the Earth sped toward the aphelion, humanity's collective nerves relaxed in tune with the Earth's increasing distance from the Sun; every new year, with the Earth falling toward the Sun, they would grow tauter with each passing day. And, every time the Earth came to the perihelion, rumors would begin to fly, proclaiming that this time the helium flash would happen. The rumors would persist until the Earth again sped toward the aphelion; then the people's fears would begin to gradually diminish, together with the shrinking of the Sun in the sky. But the next wave of fear would already be brewing.②

2. 人类的精神像在荡着一个宇宙秋千,更适当地说,在经历着一场宇宙俄罗斯轮盘赌:升上远日点和跌向太阳的过程是在转动弹仓,掠过近日点时则是扣动扳机!每扣一次时的神经比上一次更紧张,我就是在这种交替的恐惧中度过了自己的少年时代。其实仔细想想,即使在远日点,地球也未脱离太阳氦闪的威力圈,如果那时太阳爆发,地球不是被气化而是被慢慢液化,那种结果还真不如在近日点。③

It was as if humanity's spirits were caught in a cosmic swing. Or perhaps it would be better to say that we were playing a cosmic game of Russian Roulette; the

---

① 刘慈欣. 流浪地球[M]. 北京:中国科学技术出版社,2022.
② Liu Cixin. The Wandering Earth[M]. Trans. by Holger Nah. New York:Tor Trade, 2022.
③ 同①。

journeyto the aphelion and back to the perihelion was like the turning of the chamber, and passing the perihelion was just like pulling the trigger! Every time the trigger was pulled our nerves would be more frayed than the last. We passed our youth under the shadow of this oscillating terror.

Actually, when one really thought about it, the Earth never left the blast radius of the helium flash. Even at the aphelion, the only difference would have been that the Earth would have been slowly liquefied instead of being vaporized by the explosion; and that end would have been worse than what would have happened at the perihelion.[1]

第二节
练习题

一、基础练习:请翻译以下句子

1. 这场漫长的战争伴随着整个人类文明,现在仍然胜负未定,虫子并没有被灭绝,它们照样傲行于天地之间,它们的数量也并不比人类出现前少。

This long war has been going on for the entire history of human civilization. But the outcome is still in doubt. The bugs have not been eliminated. They still proudly live between the heavens and the earth, and their numbers have not diminished from the time before the appearance of the humans.

2. 一到社会上,才发现自己是个地地道道的废物,除了数学啥也不会,在复杂的人际关系中处于半睡眠状态,越混越次。

However, once I graduated and went back to the real world, I realized that I was completely useless. Other than math, I knew nothing. I was half asleep when it came to the complexities of relationships between people. The longer I worked, the worse my career.

3. 我每天夜里都在一两点才回到宿舍,听着某个室友在梦中喃喃地念着女朋友的名字,这才意识到还有另一种生活。

Returning to my dorm room at one or two in the morning and hearing a roommatemumble his girlfriend's name in his sleep was the only reminder I had of that other mode of life.

二、拓展练习:请翻译以下段落

1. 太阳的灾变将炸毁和吞没太阳系所有适合居住的类地行星,并使所有类木行星完全改变形态和轨道。自第一次氦闪后,随着重元素在太阳中心的反复聚集,太阳氦闪将在一段时间反复发生,这"一段时间"是相对于恒星演化来说的,其长度可能相

---

[1] Liu Cixin. The Wandering Earth[M]. Trans. by Holger Nah. New York: Tor Trade, 2022.

当于上千个人类历史。①

This stellar disaster would not only annihilate and consume every inhabitable telluric planet in the solar system, but it would also forever change the nature and orbits of the Jovian planets. After the primary helium flash, the heavy elements would re-accumulate in the core of the Sun and further helium flashes would repeatedly occur for a period of time. This was a "period" in the stellar sense, lasting many, many thousands of human lifetimes.②

2. 有一天,新闻报道海在融化,于是我们全家又到海边去。这是地球通过火星轨道的时候,按照这时太阳的光照量,地球的气温应该仍然是很低的,但由于地球发动机的影响,地面的气温正适宜。能不穿加热服或冷却服去地面,那感觉真令人愉快。地球发动机所在的这个半球天空还是那个样子,但到达另一个半球时,真正感到了太阳的临近:天空是明朗的纯蓝色,太阳在空中已同启航前一样明亮了。③

One day, the news reported that the ocean was thawing. When we heard it, our family again made its way to the seashore. At that time, Earth was just crossing Mars' orbit and with its approach, the strength of the Sun had again increased. It still should not have been enough to thaw the Earth on its own, but the Earth Engines ensured that the surface temperatures had reached rather pleasant heights. People everywhere were delighted that for once they did not need to wear their thermal suits.

Earth Engines still filled the sky of our hemisphere, but on the other half of the planet, people could truly feel the Sun draw closer. Their sky was bright blue and theSun was as brilliant as it had been before our exodus began. ④

第五章
第一节
练习题

一、基础练习:请翻译以下句子

1. The night wore on, cold and clear, without further alarms, and the Moon rose slowly amid equatorial constellations that no human eye would ever see.⑤
夜深了,清冷,没有其他惊扰,月亮从人类未曾目睹的赤道星座之间冉冉升起。⑥

---

① 刘慈欣. 流浪地球 [M]. 北京:中国科学技术出版社,2022.
② Liu Cixin. The Wandering Earth[M]. Trans. by Holger Nah. New York: Tor Trade, 2022.
③ 同①。
④ 同②。
⑤ Clarke C Arthur. 2001:A Space Odyssey[M]. UK: Pearson Education Ltd & Penguin Books Ltd, 1968.
⑥ 阿瑟·克拉克. 2001:太空漫步[M]. 郝明义,译. 上海:上海文艺出版社,2019.

2. In the caves, between spells of fitful dozing and fearful waiting, were being born the nightmares of generations yet to be.①

在山洞里,在时醒时睡的困乏与担惊受怕的等待中,未来世代的人才会有的梦魇,正在成形。②

3. 我没见过黑夜,我没见过星星,我没见过春天、秋天和冬天。③

I've never seen the night, nor seen a star; I've seen neither spring, nor fall, nor winter.④

4. 我出生在刹车时代结束的时候,那时地球刚刚停止转动。⑤

I was born at the end of the Reining Age, just as the Earth's rotation was coming to a final halt.⑥

二、拓展练习:请翻译以下段落

1. It whirled around, throwing its insanely daring tormentor against the wall of the cave. Yet whatever it did, it could not escape the rain of blows, inflicted on it by crude weapons wielded by clumsy but powerful hands. Its snarls ran the gamut from pain to alarm, from alarm to outright terror. The implacable hunter was now the victim, and was desperately trying to retreat.⑦

它打了个转,把这个胆大包天的加害者甩上了洞壁。然而,不论它采取什么行动,都没法躲开四面如雨而下的攻击——一双双笨拙却有力的手,舞动着一些粗糙武器而进行的攻击。它嘶吼的声音,从疼痛转为惊慌,从惊慌转为彻底的恐惧。现在,这个横行无阻的狩猎者,转而成了受害者,一心一意只想撤退。⑧

2. 比这景象更可怕的是发动机带来的酷热,户外气温高达七八十摄氏度,必须穿冷却服才能外出。在这样的气温下常常会有暴雨,而发动机光柱穿过乌云时的景象简直是一场噩梦!光柱蓝白色的强光在云中散射,变成无数种色彩组成的疯狂涌动的光晕,整个天空仿佛被白热的火山岩浆所覆盖。爷爷老糊涂了,有一次被酷热折磨得实在受不了,看到下大雨喜出望外,赤膊冲出门去,我们没来得及拦住他,外面雨点已被地球发动机超高温的等离子光柱烤热,把他身上烫脱了一层皮。⑨

Worse than the view was the scorching heat emitted by the Earth Engines.

---

① Clarke C Arthur. 2001: A Space Odyssey[M]. UK: Pearson Education Ltd & Penguin Books Ltd, 1968.

② 阿瑟·克拉克. 2001:太空漫步[M]. 郝明义,译. 上海:上海文艺出版社,2019.

③ 刘慈欣. 流浪地球[M]. 武汉:长江文艺出版社,2008.

④ Liu Cixin. The Wandering Earth[M]. Trans. by Holger Nah. New York: Tor Trade, 2022.

⑤ 同③。

⑥ 同④。

⑦ 同①。

⑧ 同②。

⑨ 同③。

Outdoors the temperature was stuck at around 160 to 180 degrees, forcing us to wear thermal suits just to leave the house. The extreme, nearly suffocating temperatures often brought torrential rains. It was always a nightmarish scene when the beam of an Earth Engine cut through dark clouds. The clouds scattered the brilliant, bluish-white light of the beam, erupting it into countless frenzied, surging halos of rainbow light that covered the entire sky like white-hot magma. One time, my senile grandfather—tormented by the unrelenting heat—couldn't take it anymore; when a heavy downpour arrived, he was so elated that he ran outside, bare to the waist. We couldn't stop him in time and the top of his skin was scalded off by the raindrops which were heated to a boil by the Earth Engines' plasma beams.①

三、思考题

（略）

第二节
练习题

一、基础练习：请翻译以下句子

1. The scene was so alien that for a moment it was almost meaningless to eyes accustomed to the colors and shapes of Earth.②

这个景象太过奇异，有那么一阵子，对已经熟悉地球上各种颜色和形状的肉眼而言，几乎是毫无意义。③

2. Far, far below lay an endless sea of mottled gold, scarred with parallel ridges that might have been the crests of gigantic waves.④

在遥远的下方，有一片无边无际、层次斑驳的金色海洋，海面散布着一道道应该是平行巨浪的波峰。⑤

3. And that golden vista could not possibly have been an ocean, for it was still high in the Jovian atmosphere.⑥

这一片金光闪闪的影像不可能是一片海洋，因为探测器还高高地位于木星的大气之中。顶多只可能是另一片云层。⑦

---

① Liu Cixin. The Wandering Earth[M]. Trans. by Holger Nah. New York: Tor Trade, 2022.
② Clarke C Arthur. 2001: A Space Odyssey[M]. UK: Pearson Education Ltd & Penguin Books Ltd, 1968.
③ 阿瑟·克拉克. 2001：太空漫步[M]. 郝明义，译. 上海：上海文艺出版社，2019.
④ 同②。
⑤ 同③。
⑥ 同②。
⑦ 同③。

二、拓展练习：请翻译以下段落

1. The night before the descent, Maya couldn't sleep. Eventually she gave up trying, and pulled herself through the rooms and corridors, up to the hub. Every object was sharp-edged with sleeplessness and adrenaline, and every familiarity of the ship was countered or overwhelmed by some alteration, a lashed-down stack of boxes or a dead-end in a tube. It was as if they had already left the Ares. She looked around at it one last time, drained of emotion. Then she pulled herself through the tight locks, into the landing vehicle she had been assigned to. Might as well wait there. She climbed into her spacesuit, feeling, as she so often did when the real moment came, that she was only going through another simulation. She wondered if she would ever escape that feeling, if being on Mars would be enough to end it. It would be worth it just for that: to make her feel real for once! She settled into her chair.①

登陆前一天，玛雅睡不着。最后她放弃在床上挣扎，站起身来到各个房间、各条走廊去看看，最后来到了中央轮轴。每一样东西看来都了无睡意，好像都打了肾上腺素；每件熟悉的东西好像都有些改变，甚至面目全非，像是一堆被撞得歪七扭八的箱子，或是一个没有出口的试管。感觉好像所有人都离开了这艘船。她看这艘船最后一眼，觉得情感已逐渐耗竭。她走到闭锁室，进入分配给她的登陆艇。也许就干脆在这里等吧。她爬进她的宇航服，感觉一下真正的时刻，但是，她还是觉得像另外一次模拟演练。单单这件事情就够了：让她真真实实感受一次！她坐回椅子上。②

2. Late that afternoon, Nadia tilted the nose of the dirigible down and circled into the wind, dropping until they were within ten meters of the ground and then releasing their anchor. The ship rose, jerked on its line, and settled downwind of the anchor, tugging at it like a fat kite. Nadia and Arkady twisted down the length of the gondola, to what Arkady called the bomb bay. Nadia lifted a windmill onto the bay's winch hook. The windmill was a little thing, a magnesium box with four vertical vanes on a rod projecting from its top. It weighed about five kilos. They closed the bay door on it, sucked out the air, and opened the bottom doors. Arkady operated the winch, looking through a low window to see what he was doing. The windmill dropped like a plumb and bumped onto hardened sand, on the southern flank of a small unnamed crater. He released the winch hook and reeled it back into the bay, and closed the bomb doors.③

---

① Robinson Stanley Kim. Red Mars[M]. New York: Bantam Books, 1993.
② 金·斯坦利·罗宾逊. 红火星[M]. 王凌霄, 译. 重庆：重庆出版社, 2016.
③ 同①。

接近傍晚的时候,娜蒂雅倾斜船首,在风中打转,在距离地面 10 米的地方放下船锚。飞船先往上扬,然后在水平线上晃了几下,在船锚的下风处停了下来,像是一只飘在空中的风筝。娜蒂雅和阿卡迪弯弯曲曲地前进,穿过长长的船舱,来到一个被阿卡迪称为投弹室的地方。娜蒂雅把风车挂在投弹室的绞盘上。风车的体积其实很小,是一个铝盒子,顶面的四个角上有四根垂直柱子,上面各有一个风车,重量大概是 5000 克。他们先把投弹门关起来,把空气抽光,再打开船底舱门。阿卡迪操作绞盘,从比较低的窗户往外看,确认风车正常着陆。风车像是一个铅锤,缓缓下降,稳稳地停在一个陨石坑南方边缘的沙地上。他放开绞盘,将绳索逐渐卷起,船底舱门跟着关上。①

第三节
练习题

一、基础练习:请翻译以下句子

1. "其实,人类把太阳同恐惧连在一起也只是这三四个世纪的事。这之前,人类是不怕太阳的,相反,太阳在他们眼中是庄严和壮美的。那时地球还在转动,人们每天都能看到日出和日落。他们对着初升的太阳欢呼,赞颂落日的美丽。"②

"Consider this," Ms. Xing told us. "We only began fearing the Sun three or four centuries ago. Before that, humanity was not afraid of the Sun. In fact, on the contrary; in their eyes, the Sun was both dignified and magnificent. Back then, the Earth still turned and people saw the Sun rise and set every single day, cheering the dawn and praising the beauty of sundown."③

2. 自第一次氦闪后,随着重元素在太阳中心的反复聚集,太阳氦闪将在一段时间反复发生,这"一段时间"是相对于恒星演化来说的,其长度可能相当于上千个人类历史。④

After the primary helium flash, the heavy elements would re-accumulate in the core of the Sun and further helium flashes would repeatedly occur for a period of time.⑤

3. 琳,你真的太聪明了,早在几年前,你就嗅出了知识界的政治风向,做出了一些超前的举动,比如你在教学中把大部分物理定律和参数都改了名字,欧姆定律改叫电阻定律,麦克斯韦方程改名成电磁方程,普朗克常数叫成了量子常数。⑥

Lin, you truly are too smart. Even a few years ago, you could feel the political

---

① 金·斯坦利·罗宾逊. 红火星[M]. 王凌霄,译. 重庆:重庆出版社,2016.
② 刘慈欣. 流浪地球[M]. 武汉:长江文艺出版社,2008.
③ Liu Cixin. The Wandering Earth[M]. Trans. by Holger Nah. New York: Tor Trade, 2022.
④ 同②.
⑤ 同③.
⑥ 刘慈欣. 三体[M]. 重庆:重庆出版社,2008.

winds shifting in academia and prepared yourself. For example, when you taught, you changed the names of many physical laws and constants: Ohm's law you called resistance law, Maxwell's equations you called electromagnetic equations, Planck's constant you called the quantum constant. ...①

二、拓展练习:请翻译以下段落

1. 沿着石阶,汪淼攀上了金字塔的顶部,看到了一处类似于古观星台的地方。平台的一角有一架数米高的天文望远镜,旁边还有几架较小型的。另一边是几台奇形怪状的仪器,很像古中国的浑天仪。最引人注目的是平台中央的一个大铜球,直径两米左右,放置在一台复杂的机器上,由许多大小不同的齿轮托举着,缓缓转动。汪淼注意到,它的转动方向和速度在不停地变化。在机器下方有一个方坑,在里面昏暗的火光中,汪淼看到几个奴隶模样的人在推动着一个转盘,为上面的机器提供动力。②

Wang climbed up the stairs and reached the apex. The platform looked like an ancient astronomical observatory. In one corner was a telescope several meters high, and next to it were a few smaller telescopes. In another corner were a few strange instruments that reminded him of ancient Chinese armillary spheres, models of objects in the sky.

His attention was drawn to the large copper sphere in the center of the platform. Two meters in diameter, it was set on top of a complex machine. Propelled by countless gears, the sphere slowly rotated. Wang noticed that the direction and speed of its rotation constantly shifted. Below the machine was a large square cavity. By the faint torchlight within, Wang saw a few slavelike figures pushing a spoked, horizontal wheel, which provided the power to the machine above.③

2. "当然,记录员站在一个底部有滑轮的架子上,位置保持在球体中心。将模拟宇宙设定到现实宇宙的某一状态后,它其后的运转将准确地模拟出未来的宇宙状态,当然也能模拟出太阳的运行状态,那名记录员将其记录下来,就形成了一本准确的万年历,这是过去上百个文明梦寐以求的东西啊。你来得正好,模拟宇宙刚刚显示,一个长达四年的恒纪元将开始,汉武帝已根据我的预测发布了浸泡诏书,让我们等着日出吧。"④

"Yes. A clerk stands on top of a shelf with a wheeled base that is kept at the

---

① Liu Cixin. The Three-body Problem[M]. Trans. by Ken Liu. New York: A Tom Doherty Associates Book, 2018.
② 刘慈欣. 三体[M]. 重庆:重庆出版社, 2008.
③ 同①。
④ 同②。

center of the sphere. After we set up the model universe to correspond to the current state of the real universe, the motion of the model thereafter should be an accurate simulation of the future, including the motion of the sun. After the clerk records the movements of the sun, we will have a precise calendar. This is the dream of hundreds of civilizations before us."

"And it looks like you have come at an opportune time. According to the model universe, a four-year-long Stable Era is about to begin. Emperor Wu of Han has just issued the order to rehydrate based on my prediction. Let's wait for sunrise!"①

第六章
第一节

练习 1

Now the spinning wheels of light began to merge, and the spokes fused into luminous bars that slowly receded into the distance, rotating on their axes as they did so. They split into pairs and the resulting sets of lines started to oscillate across one another, slowly changing their angles of intersection. Fantastic, fleeting geometrical patterns flickered in and out of existence as the glowing grids meshed and unmeshed; and the man-apes watched, mesmerized captives of the shining crystal.

They could never guess that their minds were being probed, their bodies mapped, their reactions studied, their potentials evaluated. At first, the whole tribe remained half crouching in a motionless tableau, as if frozen into stone. Then the man-ape nearest to the slab suddenly came to life.②

现在一个个旋转的光轮开始融合,轮辐也聚合成光柱。光柱一面继续沿着原来的轴线旋转,一面慢慢地后退。然后,这些旋转的光柱又各自一分为二,一分为二的光柱再开始交叉摆动,摆动中又慢慢改变交叉的角度。随着发亮网格线的结合与分离,一个个炫目的几何图案就闪耀而生,摇曳而灭。猿人呆呆地望着——在这闪烁的晶体面前,他们成了失神的俘虏。他们做梦也不会想到的是:在这段时间,他们的心智正在被探测,体态正在被记录,反应正在被研究,潜能正在被评估。起初,整个部落仿佛都冻结成石像,一动不动地半蹲在那里形成静止画面。后来,最接近巨石的那个猿人突然活了过来。③

---

① Liu Cixin, The Three-body Problem[M]. Trans. by Ken Liu. New York: A Tom Doherty Associates Book, 2018.

② Clarke C Arthur. 2001: A Space Odyssey[M]. UK: Pearson Education Ltd & Penguin Books Ltd, 1968.

③ 阿瑟·克拉克. 2001:太空漫步[M]. 郝明义,译. 上海:上海文艺出版社,2019.

练习 2

The stewardess came walking up the narrow corridor to the right of the closely spaced seats. There was a slight buoyancy about her steps, and her feet came away from the floor reluctantly as if entangled in glue. She was keeping to the bright yellow band of Velcro carpeting that ran the full length of the floor-and of the ceiling. The carpet, and the soles of her sandals, were covered with myriads of tiny hooks, so that they clung together like burrs. This trick of walking in free fall was immensely reassuring to disoriented passengers.①

空姐沿着窄窄的走道，来到右边排得很密的座位旁。她的脚步有点轻飘飘的，双脚在地毯上像是上了胶一样，勉勉强强才能抬开。沿着座船通道和船顶，全程铺着一条亮黄色的尼龙搭扣地毯，她就一直走在这条地毯上。地毯和她便鞋的鞋跟上，都布满了无数细微的小钩子，以便像芒刺一样地钩挂在一起。为了在无重力状态下走路而做的这种设计，确实可以叫晕头转向的乘客放心许多。②

练习 3

Floyd was still musing over these thoughts when his helmet speaker suddenly emitted a piercing electronic shriek, like a hideously overloaded and distorted time signal. Involuntarily, he tried to block his ears with his spacesuited hands; then he recovered and groped frantically for the gain control of his receiver. While he was still fumbling, four more of the shrieks blasted out of the ether; then there was a merciful silence.③

弗洛伊德的思绪在驰骋不已的当儿，他头盔里的扬声器突然传出一阵尖锐的电子声音，好像收音机的报时信号由于电流太强而扭曲，极其刺耳。不由自主地，他隔着航天服想用双手挡住自己的耳朵，接着他恢复镇定，拼命去摸他接收器的增益控制。在他笨拙摸索的这阵子，天外又传来四次同样尖锐的声音，然后，一切又归于静寂。④

练习 4

Despite their relative youth, Poole and Bowman were veterans of a dozen space voyages, but now they felt like novices. They were attempting something for the first time; never before had any ship traveled at such speeds, or braved so intense a gravitational field. The slightest error in navigation at this critical point and Discovery would go speeding on toward the far limits of the Solar System, beyond

---

① Clarke C Arthur. 2001: A Space Odyssey[M]. UK: Pearson Education Ltd & Penguin Books Ltd, 1968.

② 阿瑟·克拉克. 2001:太空漫步[M]. 郝明义,译. 上海:上海文艺出版社, 2019.

③ 同①。

④ 同②。

any hope of rescue. The slow minutes dragged by Jupiter was now a vertical wall of phosphorescence stretching to infinity above them—and the ship was climbing straight up its glowing face. Though they knew that they were moving far too swiftly for even Jupiter's gravity to capture them, it was hard to believe that Discovery had not become a satellite of this monstrous world.①

普尔和鲍曼虽然还都相当年轻,但已经是十来次太空之旅的老手。不过,现在这一刻,他们只觉得自己像是刚上路的菜鸟。他们在尝试的事情,前所未有。在他们之前,从没有任何宇宙飞船以这种速度航行过,也从没有挑战过如此强大的重力场。在这个关键时刻,航线上只要出一丁点错误,发现号就会一直冲向太阳系的遥远边界,再也没有任何救回的希望。时间一分一秒地缓缓而过,现在,木星成了一道垂直的磷光墙,在他们上方无穷延伸而去,而宇宙飞船则沿着这道闪闪发光的墙面,直直地往上爬。虽然他们也知道自己移动的速度其实够快,木星的重力来不及对他们产生作用,但还是很难不相信发现号已经成为这个诡异世界的一颗卫星了。②

练习 5

Those responsible for piloting the Ares pulled themselves to the control consoles and gave the orders to fire lateral control rockets. The Ares began to spin, stabilizing at four rpm. The colonists sank to the floors, and stood in a pseudogravity of 0.38 g, very close to what they would feel on Mars. Many man-years of tests had indicated that it would be a fairly healthy globe to live in, and so much healthier than weightlessness that rotating the ship had been deemed worth the trouble. And, Maya thought, it felt great. There was enough pull to make balance relatively easy, but hardly any feeling of pressure, of drag. It was the perfect equivalent of their mood; they staggered down the halls to the big dining hall in Torus D, giddy and exhilarated, walking on air.③

负责驾驶"战神号"的工作人员已经在控制台前坐定,下达指令点燃侧翼控制火箭。"战神号"开始旋转,稳定在每分钟四圈。移民从无重力的空间里掉了下来,站在0.38 g 的假重力环境中,这种感觉会很像是在火星上活动。长期的研究显示,在这样的重力环境里生活,对人体十分有益,也比一直处于无重力的舱房要好,所以这么点小麻烦是值得的。玛雅想,这种感觉真不坏。里面有足够的拉力,可以轻易地保持平衡,但是,又没有半点压力或牵绊的感觉。他们身轻如燕,一如轻快的心情,大伙儿的步履有点蹒跚,先后朝 D 舱前进,那里有丰盛的大餐等着他们。他们都很兴奋,态

---

① Clarke C Arthur. 2001: A Space Odyssey[M]. UK: Pearson Education Ltd & Penguin Books Ltd, 1968.
② 阿瑟·克拉克. 2001:太空漫步[M]. 郝明义,译. 上海:上海文艺出版社,2019.
③ Robinson Stanley Kim. Red Mars[M]. New York: Bantam Books, 1993.

度有点轻佻,仿佛身处云端。①

练习 6

She hit the ground with both feet solid, nothing tricky about it, the g familiar from nine months in the Ares; and with the suit's weight, not that much different from walking on Earth, as far as she could remember. The sky was a pink shaded with sandy tans, a color richer and more subtle than any in the photos. "Look at the sky," Ann was saying, "look at the sky." Maya was chattering away, Sax and Vlad spun like rotating statues. Nadezhda Francine Cherneshevsky took a few more steps, felt her boots crunch the surface. It was salt-hardened sand a couple of centimeters thick, which cracked when you walked on it; the geologists called it duricrust or caliche. Her boot tracks were surrounded by small systems of radial fractures. ②

她的两脚踩在地表上,踏踏实实地。在"战神号"上生活9个月之后,她已经很熟悉火星的重力了。如果记忆没有背叛她的话,穿上沉重的宇航服,跟在地球行走没有区别。天空是粉红色的,其间有黄色的阴影。这里的颜色比探测卫星传回来的照片要更明亮却也更迷离。"你看这天空。"安说,"你看这天空。"玛雅缓步离开人群,萨克斯和韦拉德在不停地旋转,像是回旋的雕像。娜蒂雅·弗朗辛·车尔尼雪夫斯基往前走了两步,感觉她的靴子在火星表面嘎吱作响。这里是盐化的沙地,大概有几厘米厚;走过之后,成团的泥沙会被踩碎。这种地形被地质学家称为铝铁硅钙壳或是钙质层。她足迹行经之地,留下一路的辐射碎片。③

练习 7

She walked along the shore of the new sea. It was at the foot of the Great Escarpment, in Tempe Terra, a lobe of ancient highlands extending into the north. Tempe had probably escaped the general stripping of the northern hemisphere by being roughly opposite the impact point of the Big Hit, which most areologists now agreed had struck near HradVallis, above Elysium. So, battered hills, overlooking an ice-covered sea. The rock looked like a red sea's surface in a wild cross chop; the ice looked like a prairie in the depths of winter. Native water, as Michel had said—there from the beginning, on the surface before. It was a hard thing to grasp. Her thoughts were scattered and confused, darting this way and that, all at the same time—it was like madness, but not. She knew the difference. The hum and keen of the wind did not speak to her in the tones of the MIT lecturer; she suffered no

---

① 金·斯坦利·罗宾逊. 红火星[M]. 王凌霄,译. 重庆:重庆出版社,2016.
② Robinson Stanley Kim. Red Mars[M]. New York: Bantam Books, 1993.
③ 同①.

choking sensations when she tried to breathe. It was not like that. Rather her thinking was accelerated, fractured, unpredictable—like that flock of birds over the ice, zigzagging across the sky in a hard wind from the west. Ah the feel of that same wind against her body, shoving at it, the new thick air like a great animal's paw... ①

她沿着新形成的海岸线走。此处位于大斜坡山脚,在坦佩台地,一片圆形古老高地往北延伸。坦佩或许因为地处撞击地点的正对面,而逃过了北半球所经历的大浩劫,大部分火星学家都同意,"大撞击"曾重创埃律西昂上方的赫拉德峡谷。因此,满目疮痍的山岭,俯视着一座冰海。岩石看起来像是海流汹涌汇聚处的一片红色海面,冰层看起来像是隆冬中的大草原。本土的水,正如米歇尔所言——从混沌之初便已存在,自古便存在于火星地表。这很难掌握。她的思绪纷乱,各种思潮同时涌现——有点像神智失常,但不是。她知道两者的区别。风的呼啸声与麻省理工学院讲师的口气完全不同,她试着呼吸时不会有喘不过气来之苦。不同的是,她的思想反倒加速了,思路涣散,无法预期——有如那群掠过冰面的飞鸟,在狂风天气里由西方成锯齿状飞过天际。噢,那种同样的风吹拂、推挤她身体的感觉,浓密的新空气像野兽的巨爪……②

练习8

While the United States was thrown into World War II in one devastating attack, China had been fighting since Japan invaded it in 1937. In a world descending into the chaos of war, Helen's mother had stood as the family's constant. By the 1940s, the Japanese held the fringes of China and were closing in. The family moved again and again, within and beyond China, trying to escape the escalating carnage. Helen's father, a general in Mao Tse-tung's Red Air Force, calculated his family's relocations based on military intelligence, but even those forewarnings couldn't keep them safe from the mounting destruction. When he moved the family to Hong Kong, his nerves calmed. The city was a safe haven under the protection of the powerful British Empire. The empire had never surrendered a colony; surely they would never submit to Japan.

These hopes shattered as the bombs rained down on the Pearl of the Orient. Helen's mother was at a neighbor's house. Trapped inside, she felt helpless to protect her children. When the thundering quieted, she raced home, shouting their names. She found Edwin and Helen clutching each other in the closet while her other two daughters crept out of their nearby hiding spot and hugged her close.

---

① Robinson Stanley Kim. Red Mars[M]. New York: Bantam Books, 1993.
② 金·斯坦利·罗宾逊. 红火星[M]. 王凌霄,译. 重庆:重庆出版社, 2016.

Helen whispered softly, her voice faltering, "We thought you were gone."①

残酷的突袭将美国卷入了第二次世界大战的旋涡,然而在太平洋另一端的中国,自 1937 年日本入侵以来,战火从来就不曾停歇。在战争带来的混乱中,海伦的母亲成了全家的中流砥柱。20 世纪 40 年代,日本已经占据了中国边境的大片领土,而且还在继续向腹地挺进。海伦一家颠沛流离,辗转中国内外,竭力试图逃脱这场不断升级的战争。海伦的父亲是一名将领,他的军事智慧为全家指明了每一次搬迁的方向,但就连父亲的洞见也无法帮助他们彻底摆脱战争的阴霾。直到一家子搬到了香港,他的心才终于放了下来。大英帝国的庇护让这座城市成了安全的天堂。这个强大的帝国不想放过任何一个殖民地,日本人的淫威也难以让他们屈服。但暴雨般的炸弹打破了东方之珠安全的幻梦。海伦的母亲被困在邻居家里,她不能离开这幢房子,但想到留在家里的孩子,她心急如焚。等到爆炸终于停歇下来,她飞奔回家,大声喊着孩子们的名字。她在衣柜里找到了挤成一团的埃德温和海伦,另外两个女儿也从角落里爬了出来,一头扑向妈妈。海伦轻声啜泣,她的声音仍在颤抖:"我们还以为你丢了。"②

练习 9

Her fingers flying across her notebook, Helen noted when the first-stage Redstone fell off and the second stage of eleven Baby Sergeants fired for six seconds before falling back to Earth. Then the third stage of three Baby Sergeants fired, sending the delicate instruments even farther up through the atmosphere. The third stage fell away, and the final stage, despite being filled with sand instead of a satellite, was speeding farther than any man-made object had ever gone. Helen's eyes popped when she calculated that it had achieved a Mach 18 velocity and climbed 3,335 miles into the air, setting a new record for altitude. With the numbers confirmed, JPL erupted in excitement, while in Alabama von Braun literally danced with joy. The launch was more successful than they had dreamed possible. Helen gleefully slipped into Chinese exclamations as both elation and weariness washed over her. They had done it, even though there was no satellite.③

海伦的手指在笔记簿上飞快地移动,她注意到第一级红石火箭分离后,第二级的 11 枚小中士点火了 6 秒钟,然后离开火箭主体坠向地球。接下来第三级的 3 枚小中士开始点火,将精密的设备送往大气层的更高处。第三级火箭分离后,第四级火箭达到了人造物品前所未有的高度,尽管它的整流罩里装的是沙子而不是卫星。根据海伦的计算,这级火箭达到了马赫数为 18 的速度,飞行轨道最高点海拔 3335 英里,新

---

① Holt Nathalia. Rise of the Rocket Girls[M]. New York: Little, Brown and Company, 2016.
② 娜塔莉娅·霍尔特. 让火箭起飞的女孩[M]. 阳曦,译. 北京:九州出版社,2022.
③ 同①.

的高度纪录就此诞生,她不由得瞪大了眼睛。这串数字得到确认以后,JPL 立即一片欢腾,远在亚拉巴马的冯·布劳恩也兴奋地跳起了舞。谁也没想到这次发射竟然如此圆满。惊喜和疲累交加之下,海伦情不自禁地喊了几句中文。他们做到了,虽然这枚火箭里没有卫星。①

练习 10

The mission's goal, as defined by President Kennedy in 1961, was to "land a man on the moon and return him safely to Earth." To accomplish this, Apollo would send a three-man spacecraft into lunar orbit. Once it was circling the moon, a second spacecraft, the lunar module, would bring two astronauts down to the surface, leaving one man in the cone-shaped command module. All three astronauts would return to Earth in the command module, parachutes slowing their descent as they splashed down into the ocean. Despite their differences, Ranger and Apollo were faces of the same coin; they were both headed to the moon, one with men aboard and one without. While Project Apollo steadily advanced, NASA worried about their inability to get a lander on the moon. They needed Ranger to work.②

正如肯尼迪总统在 1961 年说过的那样,这个任务的目标是"将人类送上月球,并把他安全带回地球"。为了达成这个目标,阿波罗计划会将一艘三人飞船送入月球轨道。接下来,第二艘飞船——登月舱——将带着两位宇航员登上月球表面,另一位宇航员则留在圆锥形的指令舱里。最后 3 位宇航员一起乘坐指令舱返回地球,降落伞会帮助他们减速,飞船最终将在大海里溅落着陆。虽然徘徊者计划和阿波罗计划有很多不同之处,但这两个项目就像硬币的两面:它们的目标都是月球,只不过一个是载人任务,另一个是无人任务。虽然阿波罗计划进展顺利,但 NASA 对登陆器能否顺利登月仍有疑虑。他们需要正常运作的徘徊者。③

第二节

练习 11

从我住的地方,可以看到几百台发动机喷出的等离子体光柱。你想象一个巨大的宫殿,有雅典卫城上的神殿那么大,殿中无数根顶天立地的巨柱,每根柱子像一根巨大的日光灯管那样发出蓝白色的强光。而你,是那巨大宫殿地板上的一个细菌,这样,你就可以想象到我所在的世界是什么样子了。其实这样描述还不是太准确,是地球发动机产生的切线推力分量刹住了地球的自转,因此地球发动机的喷射必须有一定的角度,这样天空中的那些巨型光柱是倾斜的,我们是处在一个将要倾倒的巨殿

---

① 娜塔莉娅·霍尔特. 让火箭起飞的女孩[M]. 阳曦,译. 北京:九州出版社,2022.
② Holt Nathalia. Rise of the Rocket Girls[M]. New York: Little, Brown and Company, 2016.
③ 同①。

中！南半球的人来到北半球后突然置身于这个环境中,有许多人会精神失常的。①

From my home I could see the bright plasma plumes of several hundred Earth Engines. Just imagine a gigantic palace, one as large as the Parthenon on the Acropolis. Now imagine countless colossal pillars raising from that palace, reaching to the heavens, each emitting brilliant, bluish-white light like a titanic fluorescent tube. And then there is you; you are a microbe on the palace's floor. This only begins to paint the picture of the world we lived in.

This picture, however, is not yet complete. Only the forces acting tangentially to the Earth's rotation could slow it, so the Earth Engines' jets had to be aligned to a specific angle. Those gigantic pillars of light were slanted to that angle. Now imagine what that meant for our palace, with its pillars all leaning on the very verge of toppling down! Many who came from the Southern Hemisphere went mad when suddenly confronted with this awesome vista.②

练习 12

终于,我们看到了那令人胆寒的火焰,开始时只是天水连线上的一个亮点,很快增大,渐渐显示出了圆弧的形状。这时,我感到自己的喉咙被什么东西掐住了,恐惧使我窒息,脚下的甲板仿佛突然消失,我在向海的深渊坠下去,坠下去……和我一起下坠的还有灵儿,她那蛛丝般柔弱的小身躯紧贴着我颤抖着;还有其他孩子,其他的所有人,整个世界,都在下坠。这时我又想起了那个谜语,我曾问过哲学老师,那堵墙是什么颜色的,他说应该是黑色的。我觉得不对,我想象中的死亡之墙应该是雪亮的,这就是为什么那道等离子体墙让我想起了它。这个时代,死亡不再是黑色的,它是闪电的颜色,当那最后的闪电到来时,世界将在瞬间变成蒸汽。③

Then, finally, we beheld that soul-chilling blaze. At first it was only a point of light on the horizon, but it quickly grew into an expanding arc. My breath caught in my throat as I felt myself falling into the clutches of terror. It felt as if the deck below my feet had disappeared. I imagined myself plummeting into the watery abyss below; and I fell… Ling fell with me, her wispy frame clinging to my shaking body. The other children, everyone else—the entire world—all fell. And then I remembered the riddle. When I first heard it, I had asked my philosophy teacher what color that wall was. He had told me: "It must be black." The answer had seemed off to me. I always thought that a wall of death ought to shine. That was why I had remembered it when I saw the wall of plasma. In that era, death was no

---

① 刘慈欣. 流浪地球[M]. 武汉:长江文艺出版社, 2008.
② Liu Cixin. The Wandering Earth[M]. Trans. by Holger Nah. New York: Tor Trade, 2022.
③ 同①。

longer black; it was the glare of a flash, for it would be a final flash that would vaporize the world.①

练习 13

人类的逃亡分为五步:第一步,用地球发动机使地球停止转动,使发动机喷口固定在地球运行的反方向;第二步,全功率开动地球发动机,使地球加速到逃逸速度,飞出太阳系;第三步,在外太空继续加速,飞向比邻星;第四步,在中途使地球重新自转,掉转发动机方向,开始减速;第五步,地球泊入比邻星轨道,成为这颗恒星的卫星。人们把这五步分别称为刹车时代、逃逸时代、流浪时代Ⅰ(加速)、流浪时代Ⅱ(减速)、新太阳时代。②

Humanity's exodus would proceed in five steps: First, the Earth Engines' jets would be used to counteract the Earth's movement, stopping its rotation. Second, the engines' entire power would be used to set the Earth on a new path, accelerating the Earth into escape velocity, taking it away from the Sun. Third, in outer space, the Earth would continue to accelerate as it traveled to Proxima Centauri. Fourth, in transit, the Earth Engines would be re-aligned, the Earth's rotation would be restarted and the deceleration process initiated. And then fifth, the Earth would be moored in an orbit around Proxima Centauri, becoming its planet. People also called these five steps the "Reining Age", the "Exodial Age", the "First Wandering Age" (during acceleration), the "Second Wandering Age" (during deceleration), and the "New Sun Age".③

练习 14

我们的船继续航行,到了地球黑夜的部分,在这里,阳光和地球发动机的光柱都照不到,在大西洋清凉的海风中,我们这些孩子第一次看到了星空。天啊,那是怎样的景象啊,美得让我们心醉。小星老师一手搂着我们,一手指着星空,看,孩子们,那就是半人马座,那就是比邻星,那就是我们的新家!说完她哭了起来,我们也都跟着哭了,周围的水手和船长,这些铁打的汉子也流下了眼泪。所有的人都用泪眼探望着老师指的方向,星空在泪水中扭曲抖动,唯有那个星星是不动的,那是黑夜大海狂浪中远方陆地的灯塔,那是冰雪荒原中快要冻死的孤独旅人前方隐现的火光,那是我们心中的太阳,是人类在未来一百代人的苦海中唯一的希望和支撑……④

Our ship continued its voyage, making its way into the Earth's night. Neither the light of the Sun nor the glow of the Earth Engines could be seen here. As we

---

① Liu Cixin. The Wandering Earth[M]. Trans. by Holger Nah. New York: Tor Trade, 2022.
② 刘慈欣. 流浪地球[M]. 武汉:长江文艺出版社,2008.
③ 同①。
④ 同②。

stood in the cool Atlantic breeze, we children saw our first starry sky.

God, the beauty of it was heartbreaking!

Ms. Xing arched her arm around the nearest few of us, as if to embrace us all with one hand. "Look, children," she said as she pointed to the heavens with her other hand. "There is Centaurus and that is Proxima Centauri, our new home!" Tears began trickling down her face as she spoke those words, leading her to weep.

It was an emotionally infectious moment, seeing her tears. By the time she finished, we were all sobbing. All around us—even the sailors and the captain, hardened seafarers one and all—no one could stop the tears from welling up in their eyes. Through our tears we all looked in the direction that Ms. Xing was pointing, the stars in the sky twinkling as we cried. Only one point of light did not waver; a heavenly lighthouse on the distant shores of the wild sea of the night, a faint beacon for lonely travelers freezing in the cold desolation: The star of our hearts, Proxima Centauri. It was the only hope and support for a hundred future generations, set on a course through a sea of woes.①

练习 15

我们就这样开始了地下的生活,像这样在地下 500 米处人口超过百万的城市遍布各个大陆。在这样的地下城中,我读完小学并升入中学。学校教育都集中在理工科上,艺术和哲学之类的教育已压缩到最少,人类没有这份闲心了。这是人类最忙的时代,每个人都有做不完的工作。很有意思的是,地球上所有的宗教在一夜之间消失得无影无踪。历史课还是有的,只是课本中前太阳时代的人类历史对我们就像伊甸园中的神话一样。②

Thus we began our life underground. The subterranean cities spread across the continents. They were built one-third of a mile below the surface and each one had enough space for over a million inhabitants. Under the Earth's surface, I finished primary school and entered middle school.

Most of my schooling concentrated on the physical sciences and engineering; the arts and philosophical subjects, on the other hand, were condensed to a bare minimum. The human race had no time for such distractions now. All in all, humanity was then probably the busiest it had ever been. The work never ceased and there was always more to do. Interestingly enough, the religions of the surface world vanished without a trace overnight. We still had history lessons, but our

---

① Liu Cixin. The Wandering Earth[M]. Trans. by Holger Nah. New York: Tor Trade, 2022.
② 刘慈欣. 流浪地球[M]. 武汉:长江文艺出版社,2008.

history books portrayed humanity's history under the Sun as life in a mythical paradise.①

练习 16

墨子对汪淼郑重地点点头,然后凑近他说,"知道吗,在你离开的三十六万两千年里,文明又重新启动了四次,在乱纪元和恒纪元的无规律交替中艰难地成长,最短的一次只走完了石器时代的一半,但139号文明创造了纪录,居然走到了蒸汽时代!"②

"You're right." Mozi nodded at Wang solemnly. Then he moved closer. "During the three hundred and sixty-two thousand years you've been away, civilization has been reborn four more times. These civilizations struggled to develop through the irregular alternation of Chaotic Eras and Stable Eras. The shortest-lived one got only halfway through the Stone Age, but Civilization Number 139 broke a record and developed all the way to the Steam Age."③

练习 17

"说得具体些,宇宙是一个悬浮于火海中的大空心球,球上有许多小洞和一个大洞,火海的光芒从这些洞中透进来,小洞是星星,大洞是太阳。"

"很有意思的一个模型,"汪淼看着铜球说,现在他大概能猜出那是什么了,"但其中有一个大漏洞:太阳升起和落下时,我们看到它与群星是相对运动的,而大球球壳上的所有洞孔的相对位置应该是固定的。"

"很对,所以我推出了经过修正的模型,宇宙之球是由两层球壳构成的,我们看到的天空是内层壳,外层球壳上有一个大洞,内层球壳上有大量小洞,那个外壳上的大洞透进的光在两层球壳之间的夹层反射和散射,使夹层间充满了亮光,这亮光从小洞中透进来,我们就看到了星星。"④

"Let me be more specific: The universe is a hollow sphere floating in the middle of a sea of fire. There are numerous tiny holes in the surface of the sphere, as well as a large one. The light from the sea of flames shines through these holes. The tiny ones are stars, and the large one is the sun."

"That's a very interesting model." Wang looked at the giant copper sphere again and guessed at its purpose. "But there's a problem with your theory. When the sun rises or sets, we can see its motion against the background of fixed stars. But in your hollow sphere, all the holes remain in fixed positions relative to each other."

---

① Liu Cixin. The Wandering Earth[M]. Trans. by Holger Nah. New York: Tor Trade, 2022.
② 刘慈欣. 三体[M]. 重庆:重庆出版社, 2008.
③ Liu Cixin. The Three-body Problem[M]. Trans. by Ken Liu. New York: A Tom Doherty Associates Book, 2018.
④ 同②。

"Correct! That's why I've modified my model. The universal sphere is made of two spheres, one inside the other. The sky we can see is the surface of the inner sphere. The outer sphere has one large hole while the inner sphere has many small holes. The light coming through the hole in the outer sphere is reflected and scattered many times in the space between the two spheres, filling it with light. Then the light comes inthrough the tiny holes in the inner sphere, and that's how we see the stars."①

练习 18

墨子调出了游戏界面,将时间的流逝速度稍微调快了些。一轮红日升出地平线,大地上星罗棋布的湖泊开始解冻,这些湖泊原来封冻的冰面上落满了沙尘,与大地融为一体,现在渐渐变成一个个晶莹闪亮的镜面,仿佛大地睁开了无数只眼睛。在这高处,浸泡的具体细节看不清楚,只能看到湖边的人渐渐多了起来,像春天涌出洞穴的蚁群。世界再一次复活了。②

Mozi brought up the game's interface and slightly increased the rate of passage of game time. A red sun rose above the horizon, and the numerous frozen lakes and ponds scattered over the plain began to melt. These lakes had been covered by dust and had merged into the dun ground, but now they turned into numerous mirrors, as though the earth had opened many eyes. From up so high, Wang couldn't see the details of rehydration, but he could see more and more people gathered on the shores of the lakes like swarms of ants coming out of their nests in spring. The world had once again been revived.③

练习 19

墨子从一只木箱中拿出了一块黑色圆片,"加上这片烟熏的滤镜。"说着将它插到望远镜的目镜前。汪淼将望远镜对准已升到半空的太阳,不由赞叹墨子的想象力:太阳看上去确实像一个通向无边火海的孔洞,是一个更大存在的一小部分。但进一步细看时,他发现,这个太阳与自己现实经验中的那个有些不同,它有一颗很小的核心,如果将太阳看成一只眸子,这个日核就像瞳孔。日核虽小,但明亮而致密,包裹它的外层则显得有些缺少实在感,飘忽不定,很像是气态的。而穿过那厚厚的外层能看到内部日核,也说明外层是处于透明或半透明状态的,它发出的光芒,更多的可能是日核光芒的散射。④

Mozi retrieved a black, circular piece of glass. "Use this smoked glass filter."

---

① Liu Cixin. The Three-body Problem[M]. Trans. by Ken Liu. New York: A Tom Doherty Associates Book, 2018.
② 刘慈欣. 三体[M]. 重庆:重庆出版社, 2008.
③ 同①.
④ 同②.

He inserted it in front of the eyepiece.

Wang aimed the telescope at the sun, now halfway up the sky. He was impressed by Mozi's imagination. The sun did indeed look like a hole through which a sea of fire could be seen, a small view into a much larger whole. But as he examined the image in the telescope moreclosely, he realized that the sun was different from the sun he was used to in real life. The sun here had a small core. He imagined the sun as an eye. The core was like the eye's pupil, and though it was small, it was bright and dense. The layers surrounding it, by contrast, appeared insubstantial, wispy, gaseous. The fact that he could see through the outside layers to the core indicated that those layers were transparent or translucent, and the light from those layers was likely just scattered light from the core.[①]

练习 20

那片银白色的曙光以超乎寻常的速度扩展变亮,仿佛即将升起的太阳要弥补失去的时间。很快,曙光已弥漫了半个天空,以至太阳还未升起,大地已同往日的白昼一样明亮。汪淼向曙光出现的远方看去,发现地平线发出刺眼的强光,并向上弯曲拱起,成一个横贯视野的完美弧形,他很快看出那不是地平线,是日轮的边缘,正在升起的是一颗硕大无比的太阳!眼睛适应了这强光后,地平线仍在原位显现出来,汪淼看到一缕缕黑色的东西在天边升起,在日轮明亮的背景上格外清晰,那是远方燃烧产生的烟雾。金字塔下面,一匹快马从日出方向飞驰而来,扬起的尘埃在大地上划出一道清晰的灰线,人群为其让开了一条路,汪淼听到马上的人在声嘶力竭地大喊:"脱水!脱水!!"[②]

The silvery light brightened far more rapidly than usual, as though the rising sun wanted to make up for lost time. Soon, the light covered half the sky, even though the sun was still below the horizon. The world was already as bright as midday.

Wang looked toward the horizon and saw it giving off a blinding glare. Theglowing horizon arched upward and became a curve that spread from one edge of his visual field to the other. He soon realized that he wasn't seeing the horizon, but the edge of the rising sun, an incomparably immense sun.

"After his eyes adjusted to the bright light, the horizon reappeared in its old place. Wang saw columns of black smoke rising in the distance, especially clear

---

① Liu Cixin. The Three-body Problem[M]. Trans. by Ken Liu. New York: A Tom Doherty Associates Book, 2018.

② 刘慈欣. 三体[M]. 重庆:重庆出版社, 2008.

against the glowing background of the solar disk. A fast horse rushed toward the pyramid from the direction of the rising sun, the dust from its hooves forming a distinct line across the plains."

The crowd parted before the horse, and Wang heard the rider scream at the top of his lungs: "Dehydrate! Dehydrate!"[①]

[①] Liu Cixin. The Three-body Problem[M]. Trans. by Ken Liu. New York: A Tom Doherty Associates Book, 2018.